Curious and Unusual
Civil War Stories

Monument to Major General Sterling Price at Keytesville, Missouri.

CURIOUS AND UNUSUAL CIVIL WAR STORIES

Rex T. Jackson

HERITAGE BOOKS
2017

HERITAGE BOOKS
AN IMPRINT OF HERITAGE BOOKS, INC.

Books, CDs, and more—Worldwide

For our listing of thousands of titles see our website at
www.HeritageBooks.com

Published 2017 by
HERITAGE BOOKS, INC.
Publishing Division
5810 Ruatan Street
Berwyn Heights, Md. 20740

Copyright © 2017 Rex T. Jackson

Heritage Books by the author:
A Trail of Tears: The American Indian in the Civil War
Curious and Unusual Civil War Stories
James B. Eads: The Civil War Ironclads and His Mississippi
The Sultana Saga: The Titanic of the Mississippi
Monumental Tales from the Ozarks
Timeless Stories of the West: Mountaineers, Miners, and Indians
Traces of Ozarks Past: Outlaws, Icons, and Memorable Events

Cover: "A Union ship being burnt by the Alabama" (from: *History of Our Country*, Reuben Post Halleck, American Book Company, 1923; color added) and original illustrations by the author; cover design by Debbie Riley.

All rights reserved. No part of this book may be reproduced or transmitted in any form or by any means, electronic or mechanical, including photocopying, recording or by any information storage and retrieval system without written permission from the author, except for the inclusion of brief quotations in a review.

International Standard Book Numbers
Paperbound: 978-0-7884-5780-7

Curious and Unusual
Civil War Stories

CONTENTS

1...Sutlers: Unscrupulous Swindlers
7...Gettysburg: The Ghostly Horseman at Little Round Top
13...Siege at Lexington: Battle of the Hemp Bales
23...Fort Scott: Western Frontier Post
29..."Wild Bill" Hickok: Civil War Scout and Frontier Gunfighter
37..."Kit" Carson: Civil War Soldier and Frontiersman
43...Battle of the Mules
49...Conscription Act: Rich Man's War, Poor Man's Fight
55...Sinking the *Alabama*: Daring Rover of the Confederacy
65...Commandant Henry Wirz of Andersonville Prison
73...William "Bloody Bill" Anderson: The Demise of an Infamous Guerrilla
81...Battle of Carthage: First Significant Land Battle of the Civil War
91...Fort Henry Surrenders to the Navy
97...Women in the Civil War: Fighting the Stigma of Weakness
107...Civil War Food and Rations
111...*Titanic* Survivor Wrote About the Civil War
115...Civil War Love Feast: Compassion and Mercy
119...Civil War Aeronautics: Silk Balloons
125...Losing the *Indianola*: The Audacity and Trickery of Admiral Porter
129...John Charles Fremont: "Pathfinder of the West"
137...Old Creek Chief Hopoeithleyohola: Civil War Refugees
143...Nathan Bedford Forrest: "When Churchyards Yawn"
149...Captain William Clarke Quantrill and the Desecration of His Bones
155...Colonel Samuel Colt's Multi-Shot Killing Machine
161...Additional Illustrations
173...Index
185 About the Author

Introduction

CIVIL WAR history is rich in stories about the struggle over the North's preservation of the Union and the South's preservation of the institution of slavery. Though there were multitudes of battles and skirmishes to fill history books and keep scholars occupied for years to come over the War of the Rebellion, some curious and unusual stories have been somewhat overlooked in the shuffle. In this small volume a number of lesser-known people, places, and events take "center stage" to shine under the spotlight.

For example, in the waning years of the Civil War in May 1864, Union General William Tecumseh Sherman left Chattanooga, Tennessee and marched towards Atlanta, Georgia with about 99,000 men; and by September, the Union had captured that great Southern military supply depot. As a Christmas present to President Abraham Lincoln, thanks to Gen. Sherman and his 62,000-man army's march to the sea, he was able to send a Christmas Eve telegram to him, saying: "I beg to present to you as a Christmas gift the city of Savannah [Georgia] with 150 heavy guns and plenty of ammunition and also about 25,000 bales of cotton." It must have been music to "Abe's" war-weary ears. Afterwards, Gen. Sherman headed north to join General Ulysses S. Grant to finish off the Confederacy once-and-for-all.

The war generated many strange occurrences and prompted much military experimentation with weaponry. One such invention that found a home during the War Between the States was the *CSS H.L. Hunley*—a submarine. The *"Hunley"* was the world's first underwater, combat vessel. It was shaped like a cigar and made of iron but had a ghastly reputation of transporting its brave, daring crew into eternity, including its inventor, Horace Lawson Hunley. However, the "veritable coffin," as it came to be called, did manage to sink the *USS Housatonic* on February 17, 1864, off the coast of Charleston, South Carolina, by using another experimental device—a torpedo; but after the hand-cranked "sub" delivered the weapon, it sank once again along with its doomed occupants.

Yet another curious and unusual lesser-known story involves Native Americans, who fought side-by-side with regular army troops of the North and South in Indian Territory (Oklahoma), Missouri, and Arkansas. Thousands of American Indians, mostly Cherokee, Creek, Chickasaw, Choctaw, and Seminole participated in battles at places like: Pea Ridge and Poison Spring, Arkansas; Newtonia, Missouri; and especially in the Indian Nations at Cabin Creek, Honey Springs, Fort Wayne, Round Mountain, Bird Creek, and Patriot Hills, to name a few. History buffs of the Western Theater are probably familiar with these events—but, many history books have overlooked, for some reason, the work and sacrifice of Native Americans during the Civil War.

Still another wild, unusual tale involves the theft and use of a Confederate juggernaut—an iron horse. The bold, adventurous plan was one of the most daring feats devised and undertaken behind enemy lines during the War Between the States.

About the Southern railroads, in the *Official Records of the Union and Confederate Armies* Major General George B. McClellan writing to Major General Ambrose E. Burnside on April 20, 1862, reported that the "railways in the South are represented to be in miserable condition, both as regards tracks and rolling stock...." Nevertheless, during the war, between 1861 and 1865 the system was adequate enough to transport a large amount of troops and supplies to Confederate camps and the many battlegrounds.

In the spring of 1862, Union forces of General Ulysses S. Grant and General Don Carlos Buell were advancing on Corinth, Mississippi, in order to defeat and demoralize the Confederacy in

Eastern Tennessee, with further plans to invade the nearby stronghold of Chattanooga. As a result, an ingenious plan was put into motion in March and authorized by Gen. Buell to mount a raid into the area to destroy railroad bridges from Chattanooga to Bridgeport, and thereby disrupt the flow of Southern troops and supplies. A Northern spy by the name of James J. Andrews and about a half dozen other men were sent out by Major General Ormsby M. Mitchel to do the work, but the daring scheme failed.

A second and more successful attempt was made on April 12, 1862. Andrews' plans for the mission was approved by Gen. Mitchel and undertaken by Andrews and twenty-four other raiders. The plan was to go behind enemy lines and capture a train engine and proceed to cut telegraph wires, burn railway bridges on the Georgia State and East Tennessee lines, and sabotage the track whenever possible along the way. For almost 100 miles the brave Union saboteurs created chaos, havoc, and commotion on this Great Locomotive Chase in the South.

Once the band had infiltrated the area they converged at the Marietta Hotel in Marietta, Georgia, in the Kennesaw Mountains; since some failed to make it to the rendezvous they numbered only twenty in all. When everything was ready they headed north of Marietta a short distance to Big Shanty Station and proceeded to capture their prize, a locomotive the Confederates had dubbed: *The General.*

The sky was overcast and rain continued to stubbornly plague them throughout the day. Andrews and his two engineers, Wilson W. Brown and William Knight, wasted no time uncoupling the engine, tender, and three empty boxcars from the train and climbed aboard the engine compartment; while the rest of the band scrambled onto the rear car. The engine, which was under power, was put into motion at full speed to the consternation and amazement of the Southerners that eyeballed them from the train station.

The daring group headed north toward Chattanooga, while their leader, Andrews, bravely and calmly told station officials along the route that the train was carrying a precious cargo of military supplies for General Pierre Gustave Beauregard. In *The Locomotive Chase in Georgia* found in *Battles and Leaders of the Civil War*, William Pittenger, a member of the historic party, later wrote:

"There was a wonderful exhilaration in passing swiftly by towns and country...It possessed just enough of the spice of danger—in this part of the run—to render it thoroughly enjoyable."

Meanwhile, back at the Big Shanty where it all started, an angry Confederate train conductor and foremen, William A. Fuller and Anthony Murphy, were without a train, on foot, and in hot pursuit of the stolen Confederate property. Reaching an idle handcar the two utilized it until they were eventually derailed by the destructive handiwork of the new crew aboard the *General*—steaming on ahead of them. After Fuller and Murphy recovered and finally made Etowah Station, they discovered an old engine under steam, the *Yonah*, which was being used at an iron works. Quickly, they rounded up a number of soldiers to continue the chase and, according to author Pittenger, "hurried with flying wheels towards Kingston" in the *Yonah*—while there, they commandeered a different train, the *Rome*, and continued the pursuit of the *General* with about forty armed Confederates aboard.

As Andrews' *General* steamed along they took the time to cut communication lines after passing each station along the way, damaged the track, and left debris to block their adversary's progress. Their hope was to be able to eventually set fire to railroad bridges up ahead and thereby help the Union cause in the area. Pittenger wrote: "Thus we sped on, mile after mile, in this fearful chase, around curves and past stations in seemingly endless perspective." However, fuel was running low and while their enemy continued to relentlessly steam on behind them like the "scream of a bird of prey," the decision was made to put the *General* in reverse and reluctantly abandon it and take flight to the protection of the surrounding undergrowth and wooded terrain.

In a few days, though, Andrews and all but two of his daring, Northern cohorts had been taken prisoner. Since they wore civilian clothes, however, Andrews and seven others were counted as spies, court-martialed, and sentenced to be executed; the rest of his men were never tried.

During the War Between the States many strange and unique things occurred. One such memorable event took place early in the war on April 19, 1861, in Baltimore, Maryland, commonly known as the "Baltimore Riots," where pro-Southern citizens clashed with Union troops headed for Washington "in pursuance of a call for

75,000 men made by the President of the United States." As a result of the engagement, a new piece of military technology was captured—a centrifugal steam gun dubbed the "Winans Steam Gun."

The mysterious weapon was patented by Charles S. Dickinson, but after the contraption reportedly underwent some repairs at Ross Winans' machine shop, known as "Winans & Company," the gun, for some reason, inherited his name.

Ross Winans was born in New Jersey in 1796 and became a Baltimore industrialist who built trains for the Baltimore & Ohio Railroad; among other things. Winans' secessionist beliefs, however, prompted him to support Dickinson's steam gun endeavors.

The gun weighed several tons and resembled a snowplow blade in the front with a wheel behind it that spun about 350 revolutions per minute, releasing a mini ball on each cycle. It was powered by steam and used no gunpowder—but, instead, relied upon centrifugal force to create its deadly barrage of iron balls.

Reporting about the federal troops passing through Baltimore, the Sixth Massachusetts Regiment, numbering about 800 strong, Colonel Edward F. Jones wrote on April 22, 1861, in the *Official Records of the Union and Confederate Armies*, that the "regiment will march through Baltimore in column of sections, arms at will. You will undoubtedly be insulted, abused, and, perhaps assaulted, to which you must pay no attention whatever, but march with your faces square to the front, and pay no attention to the mob, even if they throw stones, bricks, or other missiles [(a sound principle for modern-day law enforcement perhaps)]; but if you are fired upon and anyone of you is hit, your officers will order you to fire. Do not fire into any promiscuous crowds, but select any man whom you may see aiming at you, and be sure you drop him."

As expected, "...they proceeded but a short distance before they were furiously attacked by a shower of missiles, which came faster as they advanced. They increased their steps to double-quick, which seemed to infuriate the mob, as it evidently impressed the mob with the idea that the soldiers dared not fire or had no ammunition, and pistol-shots were numerously fired into the ranks, and one soldier fell dead. The order 'Fire' was given, and it was executed. In consequence, several of the mob fell, and the soldiers again advanced hastily."

As a result of the Baltimore Riot, four soldiers were killed and 36 others were wounded. As for the pro-Southern citizens, twelve lost their lives during the violent uproar. The incident caused the arrest of Ross Winans who was taken to Annapolis, Maryland, and eventually to Fort McHenry "to await the action of the civil authorities" over his loyalties and activities surrounding the steam gun, which they confiscated. In the *Official Records*, Brigadier General Benjamin F. Butler of the Massachusetts Militia, writing from Federal Hill, Baltimore, on May 15, 1861, concerning the Winans Steam Gun, reported: "I have also the honor to communicate the capture of the steam gun, and the fact that I have found men in the Sixth Massachusetts Regiment who have been able to put it in operation, and it is now in full working order."

The Union, feeling that the weapon lacked in long distance killing capabilities compared to conventional guns, abandoned it, and it disappeared.

The number of stories available for research is seemingly endless, and in the pages that follow is but a select few; however, this historical collection can go a long way to help broaden and stimulate the mind about one of the bloodiest, troubled times in American history. RTJ

Curious and Unusual Civil War Stories

Sutlers:
Unscrupulous Swindlers

DURING THE American Civil War the soldiers that managed to live through the horrors, carnage, and death of the battlegrounds, may have ended up in a field hospital where they might undergo amputations or other surgeries with no anesthesia. Under these horrifying conditions, as you might imagine, many begged for the mercy of the grim reaper to come and relieve them of their burden of life.

Between engagements, however, bivouacking soldiers often had to deal with boredom and drilling, drilling, and more drilling. They went for water, firewood, and did other necessary chores while in camp. One soldier, Captain Daniel Oakley of the 2nd Massachusetts Volunteers, who wrote about such things after the war, reported that: "Before daybreak the tramp of horses reminded us our foragers were sallying forth. The red light from the countless campfires melted away as the dawn sole over the horizon, casting its wonderful gradations of light and color over the masses of sleeping soldiers, while the smoke of burning pine-knots befogged the chilly morning air. Then the bugles broke the impressive stillness, and the roll call of drums was heard on all sides. Soon the scene was alive with blue coats and the hubbub of roll calling, cooking, and running for water to the nearest spring or stream. The surgeons looked to the sick and footsore, and weeded from the ambulances those who no longer needed to ride."

Some soldiers would while-away-the-hours by writing letters home to friends, family, and sweethearts; while others filled up their time with music, playing cards, races, and other competitions. They did whatever they could to keep their minds off of other things.

Civil War times often left many weary Blue and Gray soldiers in want of various supplies from time-to-time, and a sutler was often there with bountiful temptations to peddle to the troops, such as oysters, tobacco, coffee, sweets, pickles, sausages, tinned meats, imported sardines, boots, shoelaces, patent medicines, tailored uniforms, newspapers, cakes, fried pies, and the list goes on. These government-licensed vendors followed armies on campaign and catered to a soldier's every whim.

Soldiers browsed through clusters of tents, horse-drawn wagons, or even permanent structures when shopping with what became known as unscrupulous sutlers—a Dutch word meaning "to undertake low offices." Shoddy goods and inflated prices made many sutlers very unpopular, and midnight raids on their tents and stores became commonplace; irate customers also resorted to shoplifting to get even, or sometimes vendors were mobbed or mugged. Speaking about sutlers, a newspaper correspondent of the era wrote that they were "a wretched class of swindlers and well deserved all their troubles." One soldier reported that: "The law recognized the sutler and the orders shielded him. That was theory. Everybody kicked and cursed him and plundered him. That was practice."

Sutlers were also in danger when they set up too close to military action. Sometimes these mobile grocery stores were attacked by raiders who looted, burned, and pillaged them—stealing all manner of coveted supplies. For the most part, however, business was good and when orders came down for the troops to redeploy elsewhere, sutlers were not happy. Warren Lee Goss cried out in *Campaigning to No Purpose* that the departure of the army was "to the distress and dismay of the sutlers."

Many soldiers were drawn into debt because of the easy credit sutlers extended to their customers. Soldiers simply signed a pay voucher that vendors collected on their payment directly from the

Sutlers: Unscrupulous Swindlers

Reenactment photo taken at Carthage, Missouri, of soldiers entertaining themselves with music.

paymasters. To cash paying customers, change was sometimes given back in the form of "sutler's script," which could only be redeemed by the vendor who issued it. Concerning the sutler's system of easy credit, Private Charles F. McKenna, 155th Pennsylvania Infantry, had this to say: "Extra provisions are very dear and consequently anyone desiring to indulge in an occasional improvement on Uncle Sam's fare will soon find his pecuniary condition diminishing."

Many times, though, the thing that worried many Civil War soldiers was not the din of battle, but the conditions in camp. Many perished to disease, suicide, chronic diarrhea, exposure to the elements, and all manner of things during the long periods between conflicts. They also marched through all types of terrain in all kinds of weather, over rough and rugged roads and trails; after heavy rains the massive armies, horses, and equipment rendered the way forward muddy, foot-sucking sloppy and almost impassable to travel—but, for some, with all the hardships of war, marching, and camp-life, even those wretched sutlers might have been a welcome sight for sore eyes.

Sutlers: Unscrupulous Swindlers

Bibliography

Goss, Warren Lee, *Campaigning to No Purpose*, Battles and Leaders of the Civil War, The Century Company, 1887.

Jackson, Rex T., *Civil War Sutlers: Wretched Swindlers*, The Ozarks Reader, Vol. 2, No. 3, 2005.

Oakley, Daniel, *Marching Through Georgia and the Carolinas*, Battles and Leaders of the Civil War, The Century Company, 1887.

Rowland, Tim, *Strange and Obscure Stories of the Civil War*, Skyhorse Publishing, 2011.

Gettysburg: The Ghostly Horseman at Little Round Top

IT IS COMMONPLACE for visitors of battlefields to endure uncommon, ghostly experiences, considering the magnitude of incredible events that would have transpired on the hollowed ground. A small farming hamlet in Pennsylvania is just such a place, where on three summer days (July 1, 2, 3, 1863) the Union army consisting of about 93,000 troops faced-off with about 75,000 Confederates—the result was a harvest of death. One of the largest artillery duels of the American Civil War occurred at Gettysburg, where as many as 200 cannons were being fired simultaneously—shot and shell.

With early failures by Union generals like George McClellan, John Pope, Ambrose E. Burnside, and Joseph Hooker to defeat the stubborn Southern forces of General Robert E. Lee, Gen. Lee hoped to take advantage of this and bring the reality and carnage of war into Northern territory to prompt negotiations for an end to the bloody conflict; however, as the Army of Northern Virginia was beginning to trespass into Pennsylvania, President Abraham Lincoln made a command decision to replace Gen. Hooker with General George G. Meade to command the Army of the Potomac. Charles F. Benjamin wrote about the change in *Hooker's Appointment and Removal*: "...copies of the President's order, changing the command, were made, authenticated by the signature of the adjutant-general and addressed...to Generals Hooker and Meade. General James A. Hardie, chief of the staff of the Secretary of War, and a personal friend of both the officers concerned, was then called into the conference room and directed to start at once for Frederick City and, without disclosing his presence or business, make his way to General Meade and give him to understand that the order for him to assume the command of the army immediately was intended to be as unquestionable and peremptory as any that a soldier could receive. He was then, as the representative of the President, to take General Meade to the headquarters of General Hooker and transfer the command from the latter to the former."

The Gettysburg battleground became famous with names like Cemetery Ridge, Peach Orchard, Wheatfield, Big Round Top, Little Round Top, Seminary Ridge, Devil's Den, and so on. The land south of Gettysburg is hilly and shaped much like a fish hook; the straight part being Cemetery Ridge with the southern elevated end known as Round Top and Little Round Top; while the curved section on the northern part of the battlefield bending to the east ends at Culp's Hill. Henry J. Hunt describes much of the historic landscape in *The Second Day at Gettysburg*: "Between this wood and Plum Run is an open cleared space 300 yards wide—a continuation of the open country in front of Cemetery Ridge; Plum Run flows south-easterly toward Little Round Top, then makes a bend to the south-west, where it receives a small stream or 'branch' from Seminary Ridge. In the angle between these streams is Devil's Den, a bold, rocky height, steep on its eastern face, and prolonged as a ridge to the west. It is 500 yards due west of Little Round Top, and 100 feet lower. The northern extremity is composed of huge rocks and bowlders, forming innumerable crevices and holes, from the largest of which the hill derives its name. Plum Run valley is here marshy but strewn with similar bowlders, and the slopes of the Round Tops are covered with them...."

On the first day of battle Gen. Lee's forces managed to drive Gen. Meade's Federals from west and north to Cemetery Ridge; on the second day Confederates were victorious over a Union detachment in the Peach Orchard, but at Little Round Top it was a

Gettysburg: The Ghostly Horseman at Little Round Top

different story altogether. Colonel Joshua L. Chamberlain in command of the 20th Maine (part of the Union's Fifth Army Corps under Major General George Sykes and the First Division of Brigadier General James Barnes) had a regiment of 358 men in defense of Little Round Top; however, Col. Chamberlain sent Company B—50 men, to guard his flank which left 308 to face the coming onslaught of the 47th Alabama under Colonel James W. Jackson, Lieutenant Colonel M.J. Bulger, and Major J.M. Campbell. Col. Chamberlain and his brave Federals were outnumbered about 3 to 1. It seemed that a miracle would be needed to win the day at Little Round Top—the coveted "high ground" protecting the Union's important left flank.

The fighting at Little Round Top was fierce, and after a few hours had passed, Col. Chamberlain had only about 178 left to defend the treasured position—and, their ammunition was nearly exhausted. Each soldier carried 60 rounds of cartridges and, when it ran out, they resorted to using the cartridge-boxes leftover by their fellow comrades who no longer needed them. With little hope remaining of a victory, Col. Chamberlain gave the surprising order to "fix bayonets"—but, before his Federals had amassed for a charge, some eyewitnesses, at that pivotal moment, later recalled

Civil War reenactment photo.

seeing a mounted officer dressed in what appeared to be a Revolutionary War uniform and carrying a fiery sword. The mysterious horseman was said to be crying out for Chamberlain's surviving force to make a brave downhill dash. In *The 20th Maine at Little Round Top*, H.S. Melcher documented the action: "...the regiment leaped down the hill and closed in with the foe, whom we found behind every rock and tree. Surprised and overwhelmed, most of them threw down their arms and surrendered.

"Some fought till they were slain; the others ran 'like a herd of wild cattle....'"

Some ghost story writers contend that even the Confederates claimed to have tried to shoot the unknown, antique-looking rider, but to no avail. Afterwards, some combatants speculated that the old soldier might have been the return of George Washington who had risen from the dead to help turn the battle-tide of the ghastly War Between the States. The timely apparition could have been the encouragement they needed to make the brave, suicidal charge—or, perhaps, it was simply the leadership of Col. Chamberlain and his desperate command that brought victory at Gettysburg's Little Round Top that hot, summer day in 1863; however, it may never be known whether or not either army actually saw the spirit of George Washington.

A six gun battery.

Gettysburg: The Ghostly Horseman at Little Round Top

On the third and final bloody day Gen. Lee attempted to break through the center of the Union line, but the Federals had anticipated his move and were well prepared. In a last ditch effort of victory Gen. Lee ordered General George Picket to charge with about 14,000 men from Seminary Ridge to the wide-open, rolling Pennsylvania fields of Cemetery Ridge—they faced deadly Union rifle-fire and cannon which mowed them down like hay. The red rebel flags popped in the breeze and the sun glistened upon thousands of determined swords, rifles and their bayonets. A few lucky Southerners managed to somehow reach the hilltop and erect a Confederate flag, but it didn't fly long. What was left of Gen. Lee's Army of Northern Virginia made their way unmolested, for the most part, back to the south having lost about 28,000 men; the victorious Union army of Gen. Meade lost about 23,000 men—a total of 51,000. President Lincoln was furious that his new general in command of the Army of the Potomac did not pursue and destroy the enemy.

Upon hallowed ground the imagination can run wild. Sounds of bloodcurdling battle-cries, booming of cannons, and the crack, crack, crack of musket-fire can be heard—some swear; others claim they smell black-powder smoke or can see the clash of arms as ghostly soldiers eternally relive the action, time and again—or, just maybe, it could all be for the education of the living with the hope that its horrors will never be witnessed again.

CURIOUS AND UNUSUAL CIVIL WAR STORIES

Bibliography

Benjamin, Charles F., *Hooker's Appointment and Removal*, Battles and Leaders of the Civil War, The Century Company, 1887.

Chronicle of America, Chronicle Publications, Inc., Mount Kisco, New York, 1989.

Crain, Mary Beth, *Haunted U.S. Battlefields*, Globe Pequot Press, Guilford, Connecticut. 2008.

Halleck, Reuben Post, *History of Our Country*, American Book Company, 1936.

Hunt, Henry J., *The Second Day at Gettysburg*, Battles and Leaders of the Civil War, The Century Company, 1887.

Lengel, Edward G., *Inventing George Washington: America's Founder, in Myth and Memory*, Harper Collins Publishers, 2011.

Melcher, H. S., *The 20th Maine at Little Round Top*, Battles and Leaders of the Civil War, The Century Company, 1887.

Siege at Lexington: Battle of the Hemp Bales

IN THE Trans-Mississippi West during the American Civil War, the same hot-button issues that plagued the East spawned a multitude of battles and skirmishes which also helped to decide the fate of the nation. Much of the conflict went beyond ordinary military engagement on many occasions, and records and reports covered the action. In *The American Indian as Participant in the Civil War*, in comparing the Border War of Missouri and Kansas to Sherman's march to the sea, author Annie Heloise Abel wrote: "The irregular warfare of the border, from fifty-four [1854] on, while it may, to military history as a whole, be as important as the quarrels of kites and crows, was yet a big part of the life of the frontiersman and frightful in its possibilities. Sherman's march to the sea or through the Carolinas, disgraceful to modern civilization as each undeniably was, lacked the sickening phase, guerrilla atrocities, that made the Civil War in the West, to those at least who were in line to experience it at close range, an awful nightmare."

In many cases the war was personal and violent, as "Union and Confederate soldiers might well fraternize in eastern camps because there they so rarely had any cause for personal hostility towards each other, but not in the western." These homegrown animosities ran deep west of the Mississippi, spawning barbarous work and atrocities from the pent up hatred and wrath. One such place that witnessed its share of hostilities during the War of the Rebellion was Lexington, Missouri.

CURIOUS AND UNUSUAL CIVIL WAR STORIES

On the south bank of the Missouri River, high on a bluff overlooking the fertile river-bottoms to the north is the historic town of Lexington. First settled in the 1820s, the bustling, scenic river town quickly became a thriving community and major stopping place for traders and trappers, and steamboats and wagon trains headed west on the Santa Fe and Oregon Trails. It was a busy trade center and outfitting post and headquarters to a Westward freight firm that founded the Pony Express, as well as a popular riverboat landing; it was also a place that hosted a multi-day Civil War siege known as the Battle of Lexington.

On August 10, 1861, pro-Southern forces commanded by Missouri's ex-governor Major General Sterling Price (Old Pap), defeated Union forces at the Battle of Wilson's Creek fought near Springfield, Missouri. Soon after, Price's victorious army headed north and camped at Nevada, Mo., where they clashed on September 2, 1861, at Dry Wood Creek around Hogan's Ford about two miles south of Deerfield, Mo. The Confederates gained about 60 mules in the process as a result of the Battle of Dry Wood, which is sometimes referred to as the "Battle of the Mules."

In the *Official Records of the Union and Confederate Armies*, Union Brigadier General James H. Lane stationed at nearby Fort Scott, Kansas, reports on September 3, the day after the battle, that: "My cavalry engaged the whole force of the enemy yesterday for two hours 12 miles east of Fort Scott. It turns out to be the column of Price and [James S.] Rains, numbering from 6,000 to 10,000, with seven pieces of artillery, some 12-pounders…I am compelled to make a stand here, or give up Kansas to disgrace and destruction.

"If you do not hear from me again, you can understand I am surrounded by a superior force."

Gen. Lane was amazed that Price and his massive Southern force were content to leave the area with, for the most part, only the mules to show for it. A few days later, Price's army of about 12,000 men was advancing upon the fortified Federal outpost of Lexington, Mo.—a storehouse of arms and equipment.

The town was defended by about 2,780 Federals under the command of Colonel James A. Mulligan, who had 7 six pound guns and 2 brass mortars that heaved six-inch spherical shells. In *Battles*

Siege at Lexington: Battle of the Hemp Bales

The Greek Revival Anderson House was built of brick in 1853 by Oliver Anderson, born in Nicholasville, Kentucky in 1794. The house served as a hospital and headquarters during the Civil War and was taken and retaken several times. The house still contains bullet holes and blood stains. One cannonball is said to have entered the attic, gone through the floor and into the second floor below, where the hole in the ceiling still remains to this day.

and Biographies of Missourians by W.L. Webb, he contends that out of the 50,000 available Union troops in the state at the time, it is a mystery why the War Department left Mulligan so "unsupported." Nevertheless, Col. Mulligan made do with what he had and "constructed around the Masonic College a redan of great strength, with embrasures, parapets, and a banquette for barbette guns. The works were greatly strengthened during the five days of Price's preparation." Mulligan's men had worked diligently to shovel entrenchments on College Hill to prepare for an onslaught. Mulligan then waited "gallantly for destruction" with Colonel Peabody's command and his brave Chicago Irish Brigade; many Irish came to Chicago as immigrants as early as the 1830s and a number of them worked on the Illinois and Michigan Canal Project—at Lexington they fought for the Union army.

At first, a covered bridge defended by two six-pounders was attacked by Confederates on September 12, 1861, and in Mulligan's *The Siege of Lexington, Mo.* published in *Battles and Leaders of the Civil War* in 1887, he writes that a messenger arrived with the breaking news that the "...enemy are pushing across the bridge in overwhelming force." Price was reported to have been seen riding his horse up and down the lines personally urging his men onward. After two companies of the Missouri 13th and Company K of the Irish Brigade were sent in, the Southerners were driven back and the bridge was burned, which "...gallantly ended their work before breakfast."

Another attempt was made on the Independence Road, and again the Federals responded. This time, six companies of the Missouri 13th and the Illinois Cavalry were sent out and clashed in the Lexington Cemetery, "where the fight raged furiously over the dead."

By 3 o'clock in the afternoon, a brisk battle of artillery opened up as Price was, according to his report in the *Official Records,* "...within easy range of the college." The cannonade continued for about an hour and a half and Col. Mulligan recalled that a "lucky shot" had knocked "over the enemy's big gun, exploding a powder caisson, and otherwise doing much damage. The fight was continued until dark, and, as the moon rose, the enemy retired to

Siege at Lexington: Battle of the Hemp Bales

camp in the Fair Ground, two miles away, and Lexington was our own again."

Gen. Price recalled that after sunset, with his ammunition "nearly exhausted," and his men having gone without food for "thirty-six hours," he decided to call it a day and return to their encampment at the Fair Ground.

Pouring rain arrived on September 13, while Mulligan's Federals continued their entrenchment digging knee-deep, in some cases, wrestling in mud and rainwater while they labored. As the days progressed under the siege, Union rations depleted and the attacks continued.

Finally, after Price's Confederates had been re-supplied and reinforced, he made his final assault on the enemy's stronghold. When September 18th dawned, Price's army had swelled to about 18,000 men, and the "enemy opened," according to Mulligan, "a terrible fire with their cannon on all sides, which we answered with determination and spirit." The overwhelming army of Price "came as one dark moving mass, their guns beaming in the sun, their banners waving, and their drums beating—everywhere, as far as we could see, were men, men, men, approaching grandly."

Around noon the Southern troops had successfully taken the Oliver Anderson House, which had been confiscated by the Union for the purpose of utilizing it as a field hospital. Colonel Mulligan reported that the Confederates "filled it with their sharp-shooters, and from the scuttles in the roof, poured right into our entrenchments a deadly drift of lead."

To regain control of the house for the Union, after a company of Home Guards refused the eighty-yard suicidal task from their entrenchments, a company of the Chicago Irish Brigade was called up to do the job. "They started; at first quick, then double-quick, then on a run, then faster. Still the deadly fire poured into their ranks. But on they went; a wild line of steel, and, what is better than steel, irresistible human will." The Union controlled the Anderson House once again for about an hour or more, thanks to the Irish Brigade, but were again forced out by Price's men.

It was reported that Union troops killed three Confederate prisoners at the bottom of the grand staircase in the Anderson

House, and that a bullet hole still exists to this day in one of the staircase risers. In the *Official Records*, Price complains: "General McBride's and General Harris' divisions meanwhile gallantly stormed and occupied the bluffs immediately north of Anderson's house. The possession of these heights enabled our men to harass the enemy so greatly that, resolving to regain them, they made upon the house a successful assault, and one which would have been honorable to them had it not been accompanied by an act of savage barbarity—the cold-blooded and cowardly murder of three defenseless men, who had laid down their arms and surrendered themselves as prisoners."

Price eventually cut the Federals off from water sources "on the north, east, and south...." They suffered from its loss, their lips became cracked and their tongues swollen, and in their "agonies from thirst and their frenzy wrestling for the water [they drank the bloody water] in which the wounded had been bathed." One of Colonel Mulligan's men, commenting on the desperate water situation, also confessed that on "the morning of the 19th it rained heavily for about two hours, saturating our blankets, which we wrung out into our canteens for drinking."

As the days dragged on the attacks and siege continued and the troubles mounted for Mulligan's Federal troopers: "...the crackling of small-arms was incessant, and so thick and close were the enemy about the works, and so accurate the aim of their sharpshooters, that a man, a head, or a cap shown for a single instant above the works was sure to be saluted with fifty balls that never went many inches from the mark...The batteries were at work early and the sharpshooters occupied every tree, rock, elevation, gully, house, or other sheltering object in the vicinity of the works."

Mulligan's time was running out, and on the morning of September 20, 1861, the Confederates advanced on his thirsty force using a rather unique military tactic, by saturating hemp bales to use as a "moving fort" to advance their troops—it would go a long way in ending the historic siege. Price reported in the *Official Records* that "...continued advance of the hempen breastworks, which were as efficient as the cotton bales at New Orleans, quickly attracted the attention and excited the alarm of the enemy, who made many

Siege at Lexington: Battle of the Hemp Bales

daring attempts to drive us back."

In Mulligan's account in *Battles and Leaders* about the ingenious contribution of the mobile breastworks, it says that the "...portable hemp-bales were extended, like the wings of a partridge net, so as to cover and protect several hundred men at a time, and a most terrible and galling and deadly fire was kept up from them upon the works of the enemy...."

According to W.L. Webb, "Behind each moving bale were crouched two or three soldiers, firing as they came. Mulligan turned loose his batteries and the full tide of lead from his small-arms upon the advancing breastworks. Slowly and laboriously, but surely and steadily, the moving forts approached the Federal position." The miraculous sight of the Confederate advance using the soaked hemp bales from the river wharf below the town, must have generated a mixed bag of feelings of awe and wonder at the unfolding spectacle bearing down on them. This is the reason that many have made reference to this unusual Civil War event as the "Battle of the Hemp Bales."

With their ammunition nearly exhausted and their food and water gone, Colonel Mulligan and his brave men took a ballot and two out of six voted to fight on. However, the time had come to end the hopeless conflict and siege at Lexington and by two o'clock in the afternoon a flag of truce was carried out for their surrender. Major General Sterling Price, satisfied with the action at Lexington, said at last: "This victory has demonstrated the fitness of our citizen soldiers for the tedious operations of a siege as well as for a dashing charge."

In a despairing use of words, Mulligan knew the fight was over and said: "Our cartridges were now nearly used up, many of our brave fellows had fallen, and it was evident that the fight must soon cease...."

The battleground in and around Lexington and the Anderson House was ghastly, which left one newspaper reporter to later share what he witnessed: "Dead horses strewed the ground in every direction, producing a most intolerable odor. These, and perhaps similar circumstances, characterized the condition of affairs...and were sufficient not only to drive a man into surrender, but into suicide or insanity."

As well as capturing Lexington, the Confederates had also captured a Union riverboat steamer below the town on the Missouri River, the *Clara Bell*, which was fully loaded with stores of many and various descriptions. The victory also gained them five additional cannons, two mortars, hundreds of rifles, sabers, 750 horses and much more; and $900,000 taken by the Union was returned to the Farmers' Bank of Lexington. Union Colonel Jefferson C. Davis reporting to Major General John C. Fremont stationed in St. Louis, Missouri, writing about the Federal seizure of the money in the Lexington bank as well as their intention to secure money from other banks, like Warrensburg, wrote on September 12, 1861: "Lieutenant Pease...arrived last night with dispatches from Colonel Mulligan at Lexington...the colonel informed me that he had secured the money in the bank at that place, and was taking steps to secure that of other banks, in obedience to my orders."

A flag of truce finished the fight that was likened to and described as "a half-dozen thunderstorms had met and were battling...." Taken prisoner were "Colonels Mulligan, Marshall, Peabody, White, and Grover, Major Van Horn, and 118 other commissioned officers." When Price left Lexington his force had swelled to 22,000 men; however, by the time his victorious army had returned to southern Missouri, his force was considerably smaller.

Much of the old town of Lexington, Mo. that survived the deadly clash of Mulligan and Price, still remains, like the cannonball frozen in time at the top of one of the courthouse columns; its whizzing collision, now in deadly silence, reverberates a 19th century ordeal of chaos and conflict in the Trans-Mississippi West. Its place in American history, however, might be best summed up by a war-tale that is told of a bayoneted soldier who was said to have crawled into a closest in the Anderson House and scratched his name in blood upon the floor—wanting only to be remembered.

Siege at Lexington: Battle of the Hemp Bales

A cannonball lodged in the courthouse column at Lexington, Mo.

Bibliography

Abel, Annie Heloise, *The American Indian as Participant in the Civil War*, Arthur H. Clark Company, 1919.

Jackson, Rex T., *Lexington: A Booming River Town Under Siege*, Vol. 3, No. 1, The Ozarks Reader Magazine, Neosho, Missouri, 2006.

Lalor, Brian, *The Encyclopedia of Ireland*, Yale University Press, New Haven and London, 2003.

Mulligan, James A., *The Siege of Lexington, Mo.*, Battles and Leaders of the Civil War, The Century Company, 1887.

Official Records of the Union and Confederate Armies, Washington: Government Printing Office, 1881.

Webb, W.L., *Battles and Biographies of Missourians* or the *Civil War Period of Our State*, Hudson-Kimberly Publishing Company, Kansas City, Missouri, 1900.

Fort Scott:
Western Frontier Post

THE ROAR AND THUNDER of cannon-fire and the crackling sound of musketry, or the wagon trains and marching armies that once frequented and kept post and vigil at Fort Scott, Kansas, is not heard and is no longer needed or necessary on America's Western Frontier. The hubbub and roll call on the Parade Ground in the old fort's "Plaza," is now only a memory of wild tales reverberating from those historic times.

Fort Scott was first known as Camp Scott in honor of General Winfield Scott, but was changed in 1843. After the Indian Removal Act in the early 1800s to relocate Native Americans to Kansas Territory and the Indian Nations (Oklahoma), it was thought that a military cantonment might be prudent and should be established at a halfway point between Leavenworth, Kan. and Fort Gibson in Indian Territory. The forts along this line were intended as a series of forts on the Military Road for the purpose of protecting white settlers and further exploratory expeditions into the West. Reassigned troops from Fort Wayne, which was being built and was located just west of Maysville, Arkansas in Indian Territory, were sent to construct Fort Scott which began on May 30, 1842.

The site selected for the fort enjoyed a dry, elevated location overlooking the Military Road and near ample water supply. The Parade Ground was laid out in a square adorned with four large, two-and-a-half story, double houses for officers on the north side of the Plaza. The quarters were framed with native timber twelve

Above: Officers quarters at Fort Scott, Kansas. Below: Blockhouse at Fort Scott.

Fort Scott: Western Frontier Post

inches square and given oak flooring and walnut siding. Porches ran full length across the front, and wide stairs flowed down from above on either end of each house.

The other sides of the Plaza had troop quarters, a hospital, guard house, stables and other buildings. Water was eventually supplied by a hand-dug well about 100 feet deep. The area was entirely encompassed by a wooden stockade.

Situated on the Military Road, also known as the Fort Scott-Fort Gibson Military Road during the American Civil War, the road was a main route for all military supplies and Fort Scott was a major storehouse depot. Three block houses were added to Fort Scott in 1862-1863 during the Civil War for an added measure of protection, which were named: Fort Blair, Fort Henning, and Fort Insley. According to Wiley Britton in his book *The Civil War on the Border*, he writes that: "In addition to the rifle-pits, there were three detached bastions for the four 24-pounder siege-guns mounted at Fort Scott. On account of the large quantities of army supplies kept at the post, there were nearly always from five hundred to one thousand troops stationed there, together with some field artillery."

During the "Bleeding Kansas" era and the Border War, "Fort Scott was the ranking town among the few Federal strongholds in the middle Southwest." At election time, pro free-state and pro slave-state voters each desired to win—claiming the favor and blessing of God but keeping their rifles handy. The turmoil and trouble plaguing Kansas fueled Jayhawkers, like James Montgomery—a Christian minister who wanted to utilize places along the border of Kansas and Missouri, like Fort Scott, to defend against Border Ruffians and to launch raids of their own.

When the War Between the States erupted in 1861, Fort Scott's importance only increased. In *The American Indian as Participant in the Civil War*, author Annie Heloise Abel writes about Union General James H. Lane, and says: "He took up his position at Fort Scott, it being his conviction that, from that point and from the line of the Little Osage, the entire section of the state, inclusive of Fort Leavenworth, could best be protected."

In the *Official Records of the Union and Confederate Armies*, Gen. Lane wrote on behalf of Fort Scott to the commander of Fort

CURIOUS AND UNUSUAL CIVIL WAR STORIES

Leavenworth on August 25, 1861, begging for reinforcements and artillery, saying: "Our little force will be actively employed to defend Kansas and confuse Missourians. But, sir, I assure you that Fort Leavenworth and Kansas should be defended from this point, and the idea of holding artillery to rust at Fort Leavenworth does not strike me with any favor.

"General Weed has this moment come in from Fort Scott. He says the enemy is threatening; that a large force is marching upon us…Can you not send us re-enforcements; with it, we could play hell with Missouri in a few days."

He must have been true to his word since, O.S. Barton who penned *Three Years with Quantrell [Quantrill]* in 1914, wrote about James H. ("Jim") Lane and Colonel Charles R. Jennison that their names "became watchwords of terror to the inhabitants of the border counties of Missouri."

Fort Scott's many noteworthy contributions throughout its colorful history include: policing the Indian frontier; part of the Stephen W. Kearney Expedition of 1845; Mexican War (1846-1848); Bleeding Kansas era (1855-1861); headquarters for the Army of the Frontier in the American Civil War (1861-1865); refugee center for Native Americans; and mustered into service the brave African Americans of the 1st Kansas Colored Infantry of Civil War fame.

Its importance in American history is spotlighted and memorialized by the National Cemetery No. 1 that is located in Fort Scott, which was commissioned by President Abraham Lincoln. Present-day Fort Scott also enjoys about 14 miles of brick streets, storybook-style storefronts, and 1865-1919 Victorian mansions and architecture; however, the old Fort and Parade Ground at Fort Scott is the inspiration and reason for it all—a one-time peace-keeping outpost on the Western Frontier.

Fort Scott: Western Frontier Post

National Cemetery No. 1 commissioned by President Abraham Lincoln at Fort Scott, Kansas.

Bibliography

Abel, Annie Heloise, *The American Indian as Participant in the Civil War*, Arthur H. Clark Company, Cleveland, 1919.

Barton, Oswald Swinney, *Three Years with Quantrell*, Armstrong Herald Printing: Armstrong, Missouri, 1914.

Britton, Wiley, *The Civil War on the Border*, Vol. 2, G.P. Putnam's Sons, The Knickerbocker Press, New York and London, 1899.

Jackson, Rex T., *Old Fort Scott: On the Western Frontier*, Vol. 1, No. 1, The Ozarks Reader Magazine, Neosho, Missouri, 2004.

Official Records of the Union and Confederate Armies, Washington: Government Printing Office, 1881.

Webb, W. L., *Battles and Biographies of Missourians*, or the *Civil War Period of Our State*, Hudson-Kimberly Publishing Company, Kansas City, Missouri, 1900.

Wild Bill Hickok: Civil War Scout and Frontier Gunfighter

DURING THE 19th century Western towns lacked adequate law enforcement and city officials sought to employ marshals that were as rough and tough as the hell-raisers and gunslingers that haunted them. Many towns bustled as a result of the railroads, stagecoaches and trails heading to the West, which created the need for hotels and saloons to cater to the westward, wayward travelers; however, they also attracted their share of gamblers, outlaws, highwaymen and other riff raff that wanted to take advantage of these vulnerable communities. There was money to be had and as long as the beer flowed, the music played, the poker cards were being dealt out and the entertainment was nonstop, the need for a lawman was of the utmost importance for keeping the peace in such an active, guntotting environment.

The difficulties that awaited these brave and daring individuals willing to dawn a badge in order to keep the peace, was large. About the trouble facing town marshals, the *Daily Telegraph* in England writing about America's love affair with bearing lethal weapons wrote on October 22, 1869, about the bullies and gamblers and other such people who carry six-shooters more often than a toothpick or some other ordinary item—comparing it to what Thucydides (a Greek historian) said about the barbarism of those that carry iron; and that they hoped that the United States government would outlaw

Likeness of "Wild Bill" Hickok on the Springfield, Missouri, square.

such weapons in the future.

One fearless man that stood his ground against lawlessness and wore a badge was James Butler Hickok; better known as "Wild Bill" Hickok. Hickok was born in 1837 on a farm in Troy Grove, Illinois, in La Salle County. When he was 18-years-old he relocated to Kansas Territory where he served for a time as the constable of Monticello Township. After this he spent a few years making a living as a stagecoach driver and wagon master. When the American Civil War broke out in 1861, he served in the Union army as a scout—reportedly, in such significant conflicts as Pea Ridge, Arkansas which occurred on March 7-8, 1862, and at Newtonia, Missouri on October 28, 1864.

One year after the end of the Civil War Wild Bill Hickok was in Springfield, Mo., and according to the Springfield *Republican* dated May 28, 1898, it remembered the colorful time that Wild Bill spent in their city: "…a favorite diversion of his was to ride his horse on sidewalks and into hotels, saloons, stores and other public places. Hickok had a yellow mare named Nell which he rode at a breakneck speed through the streets whenever he chose to do so."

On the night of July 20, 1865, Wild Bill and a man named Davis K. Tutt were in the Lyons House in Springfield and were engaged in a serious hand of poker. As for Tutt, he was born in 1839 in Yellville, Arkansas, a small Ozarks town a few miles southwest of Mountain Home in Marion County, Ark. During the Civil War he served in the Confederate army in the 1st Regiment, McBride's Arkansas Infantry and as a wagon master. The fact that an ex-Union man and an ex-Southern sympathizer were playing poker together so soon after the end of the war might be some cause for alarm—nevertheless, the game ended and according to the *Republican*: "Hickok was the loser. First his money went; then his watch, a fine gold hunting-cased Waltham with a flashy chain and seal; then his diamond pin and ring."

The onetime Union scout "rose from the table completely 'strapped' and much irritated and crestfallen." There was a heated exchange of words and, being concerned about embarrassment and his manly reputation, Wild Bill fired a warning to Tutt "not to come on the street displaying the watch." To this, Tutt taunted: "I intend wearing it in the morning." Red-faced Hickok then issued his final warning that if he wore the watch in such a manner he would kill him.

The very next day on July 21, Tutt was true to his word and appeared on the square, which was known at that time as "Old Town," and walked boldly and defiantly down the west side of the Springfield square in full view of everyone—including Wild Bill, and sporting his new watch. Wild Bill stood on the corner of South Street, about seventy-five yards away and called to Tutt who was in front of the courthouse: "Dave, don't you come across here with that watch." Suddenly, in a Hollywood-style, face-to-face stand-off, Tutt reached for and drew his pistol and fired at Hickok—but missed the mark! Wild Bill, "using one arm as a rest for his Colt's Dragoon revolver," fired too, and Davis Tutt fell to the earth with a bullet-hole in his heart.

Tutt was taken inside the courthouse and died soon after. He was buried in Springfield on the west side of the Maple Park Cemetery; his tombstone is adorned, appropriately, with an illustration of a pocket watch, deck of poker cards and a pistol.

In a gun-totting environment the danger of shootouts was an ever present possibility. Even though many towns had passed ordinances against the use of or carrying dangerous weapons within the city limits, it was difficult to enforce and many ignored the law.

Wild Bill Hickok was arrested and went to trial for the killing of Davis Tutt on August 5, 1865, and was "vigorously prosecuted" by attorney Colonel Robert W. Fyan. He was defended by ex-Governor John S. Phelps and was eventually acquitted by a jury in the circuit court.

Such activity didn't slow down Hickok, the very next year he was appointed deputy United States marshal in Fort Riley, Kansas where he fought in a number of battles and skirmishes against Native Americans and served as a scout for several U.S. military officers and leaders. In 1869 he became the marshal of Hays City (present-day Hays), Kansas, and was quickly challenged by a local gunslinger by the name of Stawhan—he didn't live long. The old cow town of Hays City was notoriously rough and rugged and Wild Bill's "no weapons" law didn't make him very popular with the

Wild Bill Hickok

gun-loving, gun-packing population. While in Hays City, Hickok made many arrests—rowdy soldiers, horse thieves, the drunken and disorderly, bank robbers and so on.

About Wild Bill Hickok, the Hays City *Sentinel* February 2, 1877, reminded its readership about the time that he ruled the roost: "Bill was a quiet, peaceably disposed man—never boisterous and quarrelsome—and never starting a row. But when Bill was...convinced of an adequate cause...in a row, there was always a funeral."

After the death of Tom Smith, the marshal of Abilene, Kansas, who was killed while attempting to keep firearms and weapons outside the city limits, Will Bill was called in to establish law and order. During his nine months in 1871 as the marshal of Abilene, a way station for drovers, he would face such men as John Wesley Hardin, Ben Thompson and Phil Coe. Concerning Hardin, he was born on May 26, 1853 in Bonham, Texas and was named after the father of Methodism—his own father was a Methodist minister, but little would his parents know that John would grow up to become an infamous, Western gunfighter. He eventually was killed by John Selman in 1895. It was reported that Hardin had killed thirty people by 1874; Ben Thompson was a professional gambler and gunman who was born on November 2, 1843, in Knottingley Yorkshire, England. He served as the city marshal of Austin, Texas until he was murdered by Joe Foster in San Antonio, Texas at the Harris Variety Theater—Foster was mortally wounded in the firefight; and Phil Coe was a gambler that had an association with Ben Thompson as owner of the Bull's Head Saloon in Abilene.

One day in Abilene while Marshal Hickok was doing his job, he was confronted by Phil Coe who fired twice at Wild Bill but failed to hit him; Hickok, on the other hand, also discharged his weapon a couple of times hitting Coe in the stomach—he lingered and suffered for three days. The Abilene *Chronicle*, June 8, 1871, published Hickok's warning to the public: "...the ordinance against carrying firearms or other weapons in Abilene will be enforced. That's right. There's no bravery in carrying revolvers in a civilized community." By this time the media—authors and newspapers, were beginning to make Wild Bill Hickok a household name. He

CURIOUS AND UNUSUAL CIVIL WAR STORIES

served Abilene until December 13, 1871.

The next two years Wild Bill spent with William F. "Buffalo Bill" Cody touring eastern United States performing in a theatrical play called *Scouts of the Prairie*. Buffalo Bill, however, would go on to create the famous Buffalo Bill's Wild West Show in the 1880s where he utilized real cowboys, Western figures, and Native Americans to reenact cattle drives, historic events like the Battle of Little Big Horn and other things; they also traveled overseas and featured such renowned performers as Annie Oakley and Lakota Sioux Chief Sitting Bull. By 1909 financial problems forced a merger with Pawnee Bill, a business competitor; Buffalo Bill's Wild West Show eventually came to an end in 1915.

On March 5, 1876, Wild Bill married Agnes Thatcher Lake. Agnes Louise Messman was born in France on August 23, 1826, and relocated to Cincinnati, Ohio when she was 16-years-old. Agnes first married a circus clown, William Lake Thatcher, in 1841 and joined the Spaulding & Rogers Circus. Eventually, Agnes and her first husband formed a partnership with John Robinson and created the Robinson-Lake Circus. While the circus was at Granby, Missouri, on August 24, 1869, her husband was murdered by a man named Joe Killian. According to a newspaper report: "...a very bad element controlled matters in Newtown [Newton] County where the tragedy occurred." About two months after Wild Bill met and married the widowed Agnes he headed for South Dakota to get into gold prospecting.

The show would also end for Wild Bill Hickok on August 2, 1876, at the Number 10 Saloon in Deadwood, South Dakota. Wild Bill was playing cards in the saloon at a table with his back to the front door. Since this was not something that he would have preferred, he attempted on a couple of occasions to persuade the gambler setting across from him to trade places, but without success. As the day dragged on, however, having borrowed money from the bartender to stay in the poker game, at 4:10 p.m. a drifter by the name of Jack McCall, who had lost $10 to Hickok the day before, slipped through the saloon's doorway and quietly walked up and shot the American icon in the back of the head—the "magic" bullet passed through and out the other side and finally came to rest

Wild Bill Hickok

inside the arm of one of the other players. The hand of cards that Wild Bill held in his deadly grasp was aces and eights, which came to be known as the "Deadman's Hand."

James Butler Hickok was buried with his Sharp's rifle at the Mount Moriah Cemetery in Deadwood the next day on August 3, 1876—nearly everyone in town was in attendance to say farewell. In 1879 a cast iron fence was installed around Hickok's gravesite, thanks to the efforts of Calamity Jane who also lived in Deadwood and had cared for him. When Calamity Jane died she was buried next to Wild Bill—her dying wish. The back-shooting coward, Jack McCall, was arrested, convicted of murder and hanged for the crime.

The Old West was a place where armed individuals were commonplace, and tough, self-confident lawmen like Wild Bill Hickok—with a hint of arrogance, did their best to safeguard the weak, unarmed and innocent. As time passed and law enforcement grew, much of the "wild" was tamed-out of the West. Somewhere in the arms and embrace of this turbulent, unique American past, are held sensible lessons for our present time—as well as the future.

Bibliography

Abilene *Chronicle*, June 8, 1871.

Cheyenne *Daily News*, March 7, 1876; Galveston *News*, January 22, 1907.

Connelley, William E., *Wild Bill and His Era: The Life and Adventures of James Butler Hickok*, New York, 1933.

England *Daily Telegraph*, October 22, 1869.

Enss, Chris, *Tales Behind the Tombstones: The Deaths and Burials of the Old West's Most Nefarious Outlaws, Notorious Women, and Celebrated Lawmen*, Morris Book Publishing, 2007; *Love Lessons from the Old West: Wisdom from Wild Women*, TwoDot, 2014.

Hays City *Sentinel*, February 2, 1877.

Jackson, Rex T., *Notable Persons and Places in Missouri's History*, The Ozarks Reader, Neosho, Missouri, 2006.

McNab, Chris, *Gunfighters: The Outlaws and Their Weapons*, Thunder Bay Press, San Diego, California, 2005.

Raine, William MacLeod, *Famous Sheriffs and Western Outlaws: Incredible True Stories of Wild West Showdowns and Frontier Justice*, Skyhorse Publishing, New York, 2012.

Springfield *Republican*, May 28, 1898.

Turner, George, *Gunfighters*, Baxter Lane Company, Amarillo, Texas, 1972.

Kit Carson: Civil War Soldier and Frontiersman

LONG BEFORE the West was tamed, there were daring individuals that braved the vast wilderness of the Western Territory—mountain men, pioneers, soldiers, and pathfinders. St. Louis, Missouri was a gateway for western expansion, and American treasures and icons like Daniel Boone and his sons blazed a trail along the Missouri River to what is now Howard County where they boiled salt at a spring that was located there. The pathway to the salt lick became known as "Boone's Lick Road" or "Boone's Trace"; and the area, "Boone's Lick Country."

During this time, however, white settlers were in danger as the country was inhabited, to some degree, by hostile Indians that, for the most part, wanted to keep the lush, rich and bountiful land for themselves. Author Francis Lister Hawks who wrote *The Adventures of Daniel Boone* warned: "...the sudden attack of the Indians was like a flash of lightning, they were seen only for an instant; yet, like the lightning, they had done their work...."

Concerning the native tribes and the avarice takeover of their beloved country, Daniel Boone, the old Kentucky rifleman, wrote in his book *Daniel Boone: His Own Story* about his vision for America's future, saying that: "Here, where the hand of violence shed the blood of the innocent, where the horrid yells of savages and the groans of the distressed sounded in our ears, we now hear the praises and adorations of our Creator; where wretched wigwams stood, the miserable abodes of savages, we behold the foundations of cities laid, that, in all probability, will equal the glory of the greatest upon earth."

To further the path westward the Santa Fe Trail began in St. Louis and followed the Missouri River to Independence, Mo.—where, in the 1840s, a number of businesses had sprung up to cater to the multitudes headed westbound. The 900-mile journey across the Great Plains to Santa Fe, which was at the time only a remote outpost in New Mexico province, received caravans of pack mules and ox-drawn wagon loads of inexpensive goods made in America. About the outfitting town of Independence, in *The Oregon Trail: Sketches of Prairie and Rocky-Mountain Life*, Francis Parkman wrote that "there was an incessant hammering and banging from a dozen blacksmith's sheds, where the heavy wagons were being repaired, and the horses and oxen shod. The streets were thronged with men, horses and mules."

Along the Santa Fe Trail, however, many trading posts were established, but none more famous than Bent's Fort located on the Arkansas River in what is now present-day Colorado. The fort, named for William Bent—its cofounder, was a multiracial trading center where it traded in mostly beaver pelts and bison robes. One of the most sought after hides was the beaver, "the largest gnawing animal in North America. His body is about three feet long, and his tail nine inches. He weighs, on an average, forty pounds.

"He is a great builder—the leading carpenter among animals. He lives in and about streams of water. His house is like a huge bird's nest turned upside down. It is built in lakes or by the edge of dams and ponds, and is from eight to eighteen feet in diameter....

"The beaver uses its broad tail as a help in swimming. Its food consists of the bark of willow, poplar and birch, and the roots of the yellow pond lily. It feeds in the evening and during the night. At this time it works at house-building. Beavers are so timid and cautious that it is very difficult to watch them."

The Santa Fe Trail served from 1822 to 1880; its demise came as a result of the railroad. The town of Santa Fe was established by the Spanish in 1610. It is reported that the Palace of the Governors, an

Kit Carson

adobe structure built about 1610, is the oldest public building in the United States; also in Santa Fe is the San Miguel Church which was constructed about 1636. Santa Fe became the territorial capital in 1851; and in 1912 it became the state capital of New Mexico.

One icon of that era that made his mark in the annals of American history was "Kit" Carson who gained his Western memorable status as a frontiersman, hunter, trapper and scout. About such men, Parkman later lamented: "The mountain trapper is no more, and the grim romance of his wild, hard life is a memory of the past." To further remember those times, Parkman added: "The buffalo is gone, and of all his millions nothing is left but bones. Tame cattle and fences of barbed wire have supplanted his vast herds and boundless grazing grounds. Those discordant serenaders, the wolves that howled at evening about the traveler's camp-fire have succumbed to arsenic and hushed their savage music. The wild Indian is turned into an ugly caricature of his conqueror; and that which made him romantic, terrible, and hateful, is in large measure scourged out of him. The slow cavalcade of horsemen armed to the teeth has disappeared before parlor cars and the effeminate comforts of modern travel."

Christopher Houston Carson was born in Madison County, Kentucky on December 24, 1809. When Kit was about one year old he moved to Franklin, Missouri in Howard County; Franklin was founded in 1816 across the river from Boonville. About eight years later Kit's father perished in a fire while attempting to save the family home. Afterwards, Carson became an apprentice to a saddler but by the time he was seventeen he left Franklin to follow in the footsteps of other well-known Franklin citizens, such as Robert McKnight and William Becknell; Becknell, for example, gained his notoriety as an Indian fighter, veteran of the War of 1812, and as the first American trader to seek out Santa Fe.

Young 17-year-old Carson dreamed of becoming a mountain man in the untamed Western wilderness and, eventually, joined up with a hunting party on the Santa Fe Trail and headed into history. When he reached Taos, New Mexico, he decided to stay and for the next 14 years he devoted his life to hunting and trapping.

Carson found employment as a cook and teamster driving wagon loads of ore to Chihuahua, Mexico. He learned the Apache tongue and trapped the Gila River all the way to present-day Phoenix, Arizona and the Verde River to Flagstaff. He made hunting and trapping expeditions to California in 1829 and, in 1830 he ventured into the Rocky Mountains. Carson joined with Bent, St. Vrain & Company and hunted game for Bent's Fort in Colorado. He married an Arapaho girl, and in 1842 he returned to his Missouri homeland to bring his daughter back to stay with family.

After this, Carson befriended the "Pathfinder"—John Charles Fremont (a son-in-law of Missouri Senator Thomas Hart Benton) and became his guide for all three of Fremont's major expeditions into Oregon and California. Lieutenant Fremont and his Army Topographical Corps left St. Louis on June 10, 1842, and followed the Oregon Trail through the Great Plains and on to the Rocky Mountains. By July 10 the group had reached St. Vrain's Fort which was located between Santa Fe and Fort Laramie. About Carson, Fremont contended that he was one of the best horsemen he'd ever seen, and was the sort of man that had "a steady...eye and frank speech."

Carson's first attempt at marriage was to Waa-Nibe (Singing Grass) in 1835, who he met at the Green River Rendezvous; Waa-Nibe died about three years later due to illness. In 1841, he met and married a 17-year-old Cheyenne girl, Making-Our-Road; however, it didn't last long because Carson spent so much time away from home that Making-Our-Road decided to end the union. Then, on February 3, 1843, Carson married 15-year-old Maria Josefa Jaramillo at Our Lady of Guadalupe Church; however; in order to please her parents, Carson was, more or less, blackmailed into converting to Catholicism for her hand in marriage.

During the Mexican War from 1846-1847, Carson broke through enemy lines to secure reinforcements at San Diego; and through his many efforts he became an important part in the California conquest. After the war was over he became an Indian agent. At the outbreak of the American Civil War he was organizing soldiers to fight and, in 1863 Colonel Kit Carson entered Navajo Territory in Arizona to avenge Indian raids which had been perpetrated upon Union troops. As a result, hundreds of Navajos were captured and escorted on a cruel and brutal 450-mile trek known as the "Long Walk" to Bosque Redondo, New Mexico, to join the many Mescalero Apaches Carson had already rounded-up; during the Long Walk those who couldn't keep up were gunned down or left to die along the trail—including pregnant women. By the end of the war in 1865 Carson had attained the brevetted rank of brigadier general.

Throughout much of Carson's mountain man-style lifetime, he was known to have admitted that he "never slept under the roof of a house, or gazed upon the face of a white woman." Carson and Josefa eventually relocated to Boggsville, Colorado in 1866 where Josefa died from childbirth on April 13, 1868—just a month later, Kit Carson also died on May 23, 1868; it was three years after the end of the Civil War—he was 59-years-old.

During the Navajo War he was dubbed the "eyes of the cavalry," and was known to have captured as many as 10,000 Indians. In death, however, it was reported that he laid on a buffalo robe under the care of Doctor H.R. Tilton suffering from a ruptured abdominal aneurysm. Sometime after death, Carson and Josefa were exhumed and reburied in Taos, New Mexico at a graveyard now known as the Kit Carson Cemetery. Hundreds were in attendance at his funeral to say goodbye to the man that many considered a national treasure; however, some Native Americans may have had a different opinion of the old Indian fighter.

It didn't take long before the name of Kit Carson became a legend and a Western hero. Carson City, Nevada, founded in 1858, which became the capital of Nevada Territory in 1861 and the state capital in 1865, was named in his honor. As it turned out, the young boy that grew up in Boone's Lick Country and dreamed of becoming a mountain man had achieved his goal in life and much more—in the wilds of the American West.

CURIOUS AND UNUSUAL CIVIL WAR STORIES

Bibliography

American Heritage New Illustrated History of the United States, The Frontier, Vol. 6, Fawcett Publications, Inc., 1971.

The American Promise: A History of the United States, Bedford Books, Boston, 1998.

Boone, Daniel, *Daniel Boone: His Own Story*, Appleton, New York, 1844.

Chronicle of America, Chronicle Publications, Mount Kisco, New York, 1989.

Coit, Margaret L., *The Sweep Westward*, Time-Life Books, New York, 1963.

Enss, Chris, *Tales Behind the Tombstones: The Deaths and Burials of the Old West's Most Nefarious Outlaws, Notorious Women, and Celebrated Lawmen*, Globe Pequot Press, 2007; *Love Lessons from the Old West: Wisdom from Wild Women*, TwoDot, 2014.

George, Marian M., *Little Journey's to Alaska and Canada*, A. Flanagan Company, Chicago, 1901.

Hawks, Francis Lister, *The Adventures of Daniel Boone: The Kentucky Rifleman*, Appleton, New York, 1844.

Morgan, Robert, *Lions of the West: Heroes and Villains of the Westward Expansion*, Algonquin Books of Chapel Hill, North Carolina, 2011.

Parkman, Francis, *The Oregon Trail: Sketches of Prairie and Rocky-Mountain Life*, 1892.

Wallis, Michael, *The Wild West 365*, Abrams, New York, N.Y., 2011.

Battle of the Mules

NOT LONG AFTER the bloody Civil War engagement at Wilson's Creek on August 10, 1861, near Springfield, Missouri, Confederate Major General Sterling Price's 12,000 victorious State Guards were marching northward in the direction of Nevada, Mo. The Battle of Wilson's Creek (or Oak Hills) was described in *An Account of the Battle of Wilson's Creek* in this way: "How they did fight, these men of both armies!—fought until their gun-barrels became so hot they could scarcely hold them—fought when their leaders fell and without commands—fought when the blood and brains of their comrades were splattered into their faces—fought, many of them, until they died." Gen. Price's army made camp near Nevada on August 31, 1861.

Military campaigns required a massive amount of equipment and supplies in order to maintain a fighting force. As for the State Guards that fought under the flag of Missouri, having no government resources provided for them, it was said that, in spite of it all, they "were never demoralized by hunger." Instead, with their own hands they procured the stores of war—guns, artillery pieces, tents, foodstuff, horses, and other necessary things to continue their cause; mostly from foraging the countryside and from looting the Federals.

The very next day after Gen. Price had set up camp near Nevada, several hundred Confederate troops were deployed—initially, to reconnoiter Union strength in the Fort Scott, Kansas area; located a

Monument of Major General Sterling Price (and ex-Missouri governor) at Keytesville, Missouri.

Battle of the Mules

few miles west of Nevada. The fear that Price's overwhelming pro-Southern army might make an attack on vulnerable Fort Scott was expressed in the *Official Records of the Union and Confederate Armies* by Colonel Jefferson C. Davis, Twenty-second Indiana Volunteers, stationed at Jefferson City, Mo., to Major General John C. Fremont at St. Louis, saying: "The news...is still more convincing that Price, Parsons [Monroe M.], and Rains [James S.] are directing their movements up the Osage [River], with the view eventually,...of taking position somewhere on the river above here, probably just below Lexington [Missouri].

"Their movements certainly threaten Fort Scott, and they may attack it...."

In the meantime, Union Brigadier General James H. Lane (also known as the "Grim Chieftain") made plans to burn Fort Scott if it became necessary, to keep it from falling into Confederate hands and retreat if Colonel James Montgomery's Kansas Brigade dispatched from Fort Scott were not successful holding off Price's army. The two combative forces clashed on September 2, 1861, along the banks of Big Dry Wood Creek around Hogan's Ford about two miles south of present-day Deerfield, Mo.

In the *Official Records* Brig. Gen. Lane writes about the battle and says: "My cavalry engaged the whole force of the enemy yesterday for two hours 12 miles east of Fort Scott." Lane estimated Price's troop strength to be about 6,000 to 10,000 with seven cannons. Lane wrote that after the battle he left his "cavalry to amuse the enemy until we could establish ourselves here and remove our good stores from Fort Scott. I am compelled to make a stand here, or give up Kansas to disgrace and destruction."

"The cavalry we engaged are armed with minie rifles," and "their artillery are some of the guns taken from our army at the battle near Springfield," facts reiterated by Lane of how Price's army oftentimes obtained equipment and needed supplies. Gen. Lane also reported that he was protecting Fort Scott with about 800 regular and irregular cavalry troops; as well as Barnesville (also known as Fort Lincoln) about 12 miles northeast of Fort Scott with about 250 men, stationed in "log buildings" and having "earth entrenchments."

At Barnesville, Gen. Lane cried for reinforcements to defend

Kansas from Price's threatening Missouri State Guards; however, Lane's call for additional aid was unnecessary as Price and his superior force was content to leave the area without invading Kansas or capturing Fort Scott for the Confederacy.

After the battle-cries and the thunderous duel of cannon and musket-fire at Dry Wood had fallen silent and the smoke-filled air of the battleground had dissipated—ironically, the only thing the Confederates had gained as a result of the conflict was about 60 mules. Needless-to-say, Gen. Lane was amazed that Price was content to leave the area with only the mules to show for it. Afterwards, some began to refer to the hotly contested skirmish as the "Battle of the Mules."

Gen. Lane in the *Official Records* wrote: "I cannot believe, however, that that army has retreated satisfied with the stealing of 60 mules and with a loss of from 150 to 200 men in killed and wounded."

By mid-September the Missouri State Guards would leave southwest Missouri and travel north to renew the fight at Lexington, located on and overlooking the Missouri River. The unusual battle that added a number of mules to Price's war effort would soon be replaced by another one that would come to be known as the "Battle of the Hemp Bales."

Bibliography

Holcombe & Adams, *An Account of the Battle of Wilson's Creek*, Dow & Adams, 1883.

Official Records of the Union and Confederate Armies, Washington: Government Printing Office, 1881.

Webb, W. L., *Battles and Biographies of Missourians*, or the *Civil War Period of Our State*, Hudson-Kimberly Publishing Company, 1900.

Conscription Act: Rich Man's War, Poor Man's Fight

AS THE CIVIL WAR dragged on and the death toll and battleground carnage continued to pile up, the once popular, glorious ideal of enlisting to fight in the war began to wane. At first, President Abraham Lincoln and Confederate President Jefferson Davis had an easy-time-of-it calling for volunteers who were eager to serve and fight for their governments—patriotically; however, the inconceivable news that kept on pouring in from both east and west of the Mississippi River, dried up the enthusiastic, flowing stream of volunteers and caused a rash of desertions—and so, as a result, something had to be done to fill the ranks. The solution first came to the South in 1862, and finally to the North on March 3, 1863, by way of the Conscription Act—compulsory (forced) military service for the first time in the history of the United States.

President Lincoln's executive order passed by the Congress allowed the War Department to draft men from 18 to 45 years of age. The "Act" had to be enforced by federal marshals, due to the unpopularity of the new law. Many believed that conscription was unfair, prejudice, and promoted favoritism. One provision in the draft law allowed for draftees to be able to hire a substitute to take their place by simply paying a $300 fee; at that time, $300 was

CURIOUS AND UNUSUAL CIVIL WAR STORIES

Abraham Lincoln Statue

"Statue by St. Gaudens at Lincoln Park, Chicago. A replica of this statue was placed in front of the Houses of Parliament in London, England, in 1921."

(From: *History of Our Country*, Reuben Post Halleck, American Book Company, 1923.)

Conscription Act

about a year's income for the unskilled laborer. The money raised was to help the war effort, but many believed that it favored the rich and left the poor to carry the deadly load. Democrats who, for the most part, voted against conscription, dubbed it "aristocratic legislation," while many common people both in the North and South cried: "A rich man's war and a poor man's fight."

One practice that arose as a result of the draft law was known as "bounty-jumping," where an enlistee would claim his bounty, desert, and then reenlist somewhere else under a different name and again collect the army's cash bounty. One man, who eventually went to prison for his antics, claimed that he had "jumped" 32 times.

In a *History of Our Country*, author Reuben Halleck wrote that: "At the North a conscripted man was allowed to escape service by furnishing money enough to secure a substitute. Men would sometimes accept this money and contrive to escape either on the way to the South or during a battle...."

Many of the substitutes were taken from the recent immigrant population that had not yet applied for American citizenship and were not yet eligible for conscription. Local officials were also more inclined to select for military service the poor and the immigrants, rather than draft the well-to-do—before long, tempers flared and lawless Pandemonium followed; as well as draft dodging and rampant desertion.

On July 13-16, 1863, some of the worst riots in U.S. history erupted in New York City's East-Side, where many foreign-born citizens resided—mostly poor Irish workers. Many protested while other angry men descended upon the draft-drawing station, setting it ablaze; after this, they challenged police officers who opened fire on them; still others resorted to arson, looting, and lynching. The history-making riot was sometimes referred to as the "bloody week."

The anti-draft rioters singled out African Americans because they believed that they would migrate northward in droves and jeopardize their livelihoods by gobbling up housing and all the available jobs. The immigrant workers lived in low-class dwellings, suffered from inflation, and did not want to fight to end slavery. As a result, they killed and also hanged two African Americans and burned down the Colored Orphan Asylum. They also attacked the homes of the rich, ransacked stores and the pro-war New York *Tribune* newspaper, and did many other lawless things as well. Some reports said that as many as 70,000 may have participated in the reign of terror and madness.

In order to stop the violence, five Union regiments that had fought at the Battle of Gettysburg a few days before, were called in with artillery and Gatling guns; along with naval and state militia troops. Before it was over, 105 souls had been taken.

The Emancipation Proclamation, ironically, brought about 180,000 African Americans into the ranks—which, despite the racist anger leveled upon them, ended up helping to quell the nation's crisis of its depleted military forces. After the "Proclamation" it revealed that the War Between the States was to, not only preserve the Union, but to end human slavery, which helped to gain more favor overseas. About the cost of the war and the loss of about 700,000 men, Helleck wrote: "It would have been cheaper to free the slaves by paying full value for them. War has proved to be a wasteful, as well as a most cruel, way of settling a dispute...the United States did what other civilized nations had already done, although on a smaller scale, without war."

The Conscription Act was disliked in many places and by many citizens, and tensions ran high. To some Americans, considering the high battlefield death toll, the cost of a man's life came to $300—a lot of money to a poor person. It still is.

Bibliography

The American Promise: A History of the United States, Bedford Books, Boston, 1998.

Faragher, John Mack, *Out of Many: A History of the American People*, Prentice Hall, Upper Saddle River, New Jersey, 1994.

Goldfield, Davis R., *The American Journey: A History of the United States*, Prentice Hall, Upper Saddle River, New Jersey, 1998.

Guitteau, William Backus, *The History of the United States*, Hougton Mifflin Company, 1942.

Halleck, Reuben Post, *History of Our Country*, American Book Company, 1923.

Legrand, Jacques, *Chronicle of America*, Chronicle Publications, Inc., Mount Kisco, New York, 1989.

Sinking the *Alabama*: Daring Rover of the Confederacy

CIVIL WAR BATTLES occurred in many places both east and west of the Mississippi River, on inland waterways, in the Gulf of Mexico, and along the coastlines of the United States—but, oddly enough, it didn't stop there; conflicts erupted in the open seas, as well. One unusual, lesser-known battle, however, was waged in European waters in the English Channel near Cherbourg, France, between the warship *USS Kearsarge* and the notorious man-of-war *CSS Alabama*.

Cherbourg, located at the mouth of the Divette River on the English Channel, served as a seaport for ocean-going vessels. About 190 miles northwest of Paris, the city of Cherbourg and its harbor is protected by a breakwater (offshore structure) about 2 miles long—making it suitable for large ships to come and go. The geographical location on France's northern coastline made Cherbourg a perfect place to dock or disembark to destinations across the Atlantic Ocean. In 1912, the grand ocean liner *Titanic*, on its maiden voyage, dropped anchor at Cherbourg for an hour and a half to pick-up passengers before it steamed off into the night on its way to tragic history.

On June 11, 1864, the *Alabama* commanded by Confederate Captain Raphael Semmes, limped into Cherbourg in need of repairs; just how many of these repairs were actually completed is uncertain. Capt. Semmes recalled that: "Our little ship was now showing signs of the active work she had been doing. Her boilers were burned out, and her machinery was sadly in want of repair. She was loose at every joint, her seams were open, and the copper on her bottom was in rolls. We therefore set our course for Europe...entered the port of Cherbourg, and applied for permission to go into dock."

The *Alabama*, the infamous Confederate raider-cruiser, was built by the Lairds of Birkenhead, England, for the Confederate States of America. One of the senior partners involved with its construction, confessed to the House of Commons "that she left Liverpool a perfectly legitimate transaction." Back in America, the Union may have had a different opinion of the lucrative deal. In order to make the whole package look politically neutral, the *Alabama* set off on a test run with ladies and gentlemen aboard as a "ruse." Once out in the channel the new ship was met by a tug boat which removed the not-so-innocent passengers and, upon reaching the island of Terceira in the Azores, the *Alabama* received its armament.

The *CSS Alabama*, with its two 300-horsepower horizontal steam engines; barkentine rigged wooden propellers; overall length of 220-feet; beam, 32-feet; depth, 17-feet; able to carry 350 tons of coal; received 8 forward guns; one rifled 100-pound Blakely; one 8-inch solid-shot gun; and six broadside 32-pounders. The *Alabama's* crew consisted of 120 sailors and 24 officers when it set sail, and for almost two years lurked the seas terrorizing Union merchant ships to disrupt Northern commerce.

Meanwhile, the Union's *USS Kearsarge*, a two-engine sloop anchored on the Scheldt River off Flushing, Holland, in the Netherlands, with a crew of 163 sailors and officers, under the command of Captain John Aucrum Winslow, received word that the *Alabama* was at Cherbourg; they set sail at once.

The *Kearsarge* was a warship that had equal firepower to the *Alabama*; however, one possible difference which was unknown to Capt. Semmes was that the *Kearsarge* was chain-clad—wearing an armor of heavy chain and a veneer of hardwood on its flank.

When the *Kearsarge* arrived at Cherbourg they could see the Confederate flag boldly and defiantly flapping in the breeze above the *Alabama* within the breakwater. Word had spread like a wildfire throughout the region that an American Civil War battle was

Sinking the Alabama

A Union ship being burnt by the *Alabama*.

(From: *History of Our Country*, Reuben Post Halleck, American Book Company, 1923.)

imminent, and so many spectators hoping to get a front-row seat to a spectacular fight, began to assemble—as many as 15,000. The day was perfect for the sightseers to witness this history-making sea-fight. Hundreds of people arrived from Paris on excursion trains, while others from Cherbourg and the surrounding countryside thronged the cliffs and shorelines—some had even been given spyglasses and camping stools to enjoy the dueling ships. Reportedly, the famous painter, Edouard Manet, was also in attendance; he would later go on to paint the memorable battle-scene. The crowd was now set to experience the roar and thunderous booming of cannon-fire in thick clouds of gunpowder smoke—and death.

On June 19, 1864, with a challenge having been made and accepted, Capt. Winslow and his brave Unionists had "determined not to surrender, but fight until the last, and, if need be, to go down with flying colors." The *Alabama* was also ready for a fight, executive officer John McIntosh Kell recalled: "As we rounded the breakwater we discovered the *Kearsarge* about seven miles to the northward and eastward," and they headed for her.

Capt. Semmes climbed up upon a gun-carriage and calling his sailors to him, addressed them in this way: "Officers and Seaman of the 'Alabama': You have at length another opportunity of meeting the enemy—the first that has been presented to you since you sank the *Hatteras*! In the meantime you have been all over the world, and it is not too much to say that you have destroyed, and driven for protection under neutral flags, one-half of the enemy's commerce, which at the beginning of the war covered every sea. This is an achievement of which you may well be proud, and a grateful country will not be unmindful of it. The name of your ship has become a household word wherever civilization extends! Shall that name be tarnished by defeat? The thing is impossible! Remember that you are in the English Channel, the theater of so much of the naval glory of our race, and that the eyes of all Europe are at this moment upon you. The flag that floats over you is that of a young Republic, which bids defiance to her enemy's whenever found! Show the world that you know how to uphold it! Go to your quarters."

Sinking the Alabama

The day being Sunday, the crew of the *Kearsarge* were dressed in their finest and set to attend the "divine service." As the bell for service began to sound, someone shouted: "She's coming, and heading straight for us!" The drumbeat for general quarters being made, Capt. Winslow suddenly decided it was time to put down his prayer-book, abandon the Sunday service, and get to the bloody work at hand.

By this time the two ships were at top speed, and at a distance of about one mile the *Alabama* opened up with its 100-pound pivot-gun on the starboard bow. Before long, they were pounding each other with their broadside guns at about 500 yards.

John M. Browne, the surgeon aboard the *Kearsarge*, later reported: "The firing of the *Alabama* was rapid and wild, getting better near the close; that of the *Kearsarge* was deliberate, accurate, and almost from the beginning productive of dismay, destruction, and death."

In the official report of Capt. Semmes, he remembered the horrors of it all, saying: "The firing now became very hot, and the enemy's shot and shell soon began to tell upon our hull, knocking down, killing, and disabling a number of men in different parts of the ship."

Kell, aboard the *Alabama*, also recalled the damage and carnage being delivered by the Union warship's 11-inch shells upon their quarter-deck: "Our decks were now covered with the dead and wounded, and the ship was careening heavily to the starboard from the effects of the shot-holes on her water-line...The port side of the quarter-deck was so encumbered with the mangled trunks of the dead that I had to have them thrown overboard, in order to fight the after pivot-gun."

Such realities of war, by this time, would have surely sickened the thousands of openmouthed sightseers gawking and gaping from the coast of Cherbourg; however, afterwards in Paris it became the talk-of-the-town. Back in the United States, though, the Civil War battle in the English Channel at Cherbourg received far less national publicity than did the headlines and interest generated by the history-making clash of the Union ironclad *Monitor* and the Confederate ironclad *Virginia* (or *Merrimac*).

CURIOUS AND UNUSUAL CIVIL WAR STORIES

They came to blows on March 9, 1862, at Hampton Roads, Virginia. "The wooden hulled *Merrimac* layered with two-inch-thick iron plating and armed with ten guns, engaged the smaller but more radical designed *Monitor*...with its armament and revolving turret equipped with two eleven-inch guns." The battle was considered a draw. About the significance of the engagement between the new ironclads, in *The First Fight of Iron-clads*, John Wood believed that: "...in some respects the most momentous naval conflict ever witnessed. No battle was ever more widely discussed or produced a greater sensation. It revolutionized the navies of the world...Rams and iron-clads were in future to decide all naval warfare. In this battle old things passed away, and the experience of a thousand years of battle and breeze was forgotten."

Even though the *Alabama* was in trouble and was slowed in her response, she continued to dish out a deadly dose of broadside cannon-fire. After making seven rotations on the circular battle-track with the *Kearsarge*, the *Alabama*, settling, began to attempt escape as it started into the eighth rotation.

According to surgeon Browne aboard the *Kearsarge* "...a few well-directed shots hastened the sinking. Then the *Alabama* was at our mercy. Her colors were struck, and the *Kearsarge* ceased firing."

The *Alabama* was going down to Davy Jones' Locker, so a white flag was ordered over the stern and every man was told to abandon ship. Kell of the *Alabama* reported that with her bows high up into the air the ship was "graceful even in her death-struggle, she in a moment disappeared from the face of the waters. The sea now presented a mass of living heads, striving for their lives." It sank in forty-five fathoms of water about four and a half miles from the west side of Cherbourg's breakwater. "Thus sank the terror of merchantmen, riddled through and through...."

Not far away, the steam-yacht *Deerhound* which belonged to John Lancaster of Lancashire, England, steamed to the rescue and plucked a number of officers and men from the water; a French pilot-boat and other boats from the *Kearsarge* eventually arrived on the scene to lend a hand, as well.

Some dissatisfaction later surfaced when Capt. Semmes

Sinking the Alabama

complained and even brought religion into the fight, saying: "Although we were now but four hundred yards from each other, the enemy [*Kearsarge*] fired upon me five times after my colors had been struck. It is charitable to suppose that a ship-of-war of a Christian nation could not have done this intentionally."

In rebuttal, Browne also injects religion into the battle, contending that: "He [Capt. Semmes] is silent as to the renewal by the *Alabama* of the fight after his surrender—an act which, in Christian warfare, would have justified the *Kearsarge* in continuing the fire until the *Alabama* had sank beneath the waters."

The so-called "Christian warfare" lasted just over an hour, and took many brave sailors to the bottom of the sea; and many others were seriously wounded. The daring rover of the Confederacy was gone forever, lost in historic waters over 5,000 miles from the Confederate capital of Richmond, Virginia.

The *Monitor* and the *Merrimac* (or *Virginia*).

(From: *History of Our Country*. Rueben Post Halleck, American Book Company, 1923.)

Bibliography

Browne, John M., *The Duel Between the "Alabama" and the "Kearsarge"*, Battles and Leaders of the Civil War, The Century Company, 1887.

Funk & Wagnalls New Encyclopedia, Funk & Wagnalls, Inc., New York, 1979.

Jackson, Rex T., *James B. Eads: The Civil War Ironclads and His Mississippi*, Heritage Books, Inc., Bowie, Maryland, 2004.

Kell, John McIntosh, *Cruise and Combats of the "Alabama"*, Battles and Leaders of the Civil War, The Century Company, 1887.

Legrand, Jacques, *Chronicle of America*, Chronicle Publications, Inc., 1989.

Marriott, Leo, *Titanic*, Smithmark, 1997.

Rowland, Tim, *Strange and Obscure Stories of the Civil War*, Skyhorse Publishing, 2011.

Wood, John Taylor, *The First Fight of Iron-clads*, Battles and Leaders of the Civil War, The Century Company, 1887.

Commandant Henry Wirz of Andersonville Prison

MAN'S INHUMANITY to man rears its ugly head on occasion and stains and soils the history of civilization. It seems to respect no boundaries, and is best, though appalling, to be remembered. The past is riddled with occurrences that many would rather forget, and a perfect example of this is what happened at a place called Andersonville; a notorious Civil War prison camp in the South.

Andersonville Prison is often compared to other infamous prison camps like Auschwitz, which was established under Germany's dictator Adolf Hitler near Cracow, Poland, in 1940. Hitler created a number of these concentration camps in occupied countries during WWII where he imprisoned millions of Russians, gypsies, and about every Jew in Europe, among others. In Hitler's camps, Germans somehow tolerated and exercised heinous and unimaginable atrocities, as they tortured, starved, gassed, and gunned down helpless men, women, and children. In Auschwitz alone, more than 4,000,000 souls were brutally taken. At Buchenwald Prison Camp, for example, German captors collected and tanned human skin to make book covers, lamp shades and other things; and monstrous experimentations on human subjects were also perpetrated upon the defenseless.

After Germany and Japan were defeated, those responsible for such concentration-camp crimes of war were tried by United Nation tribunals and, when found guilty, they were hanged or incarcerated.

In the fall of 1863, however, Confederate Brigadier General John

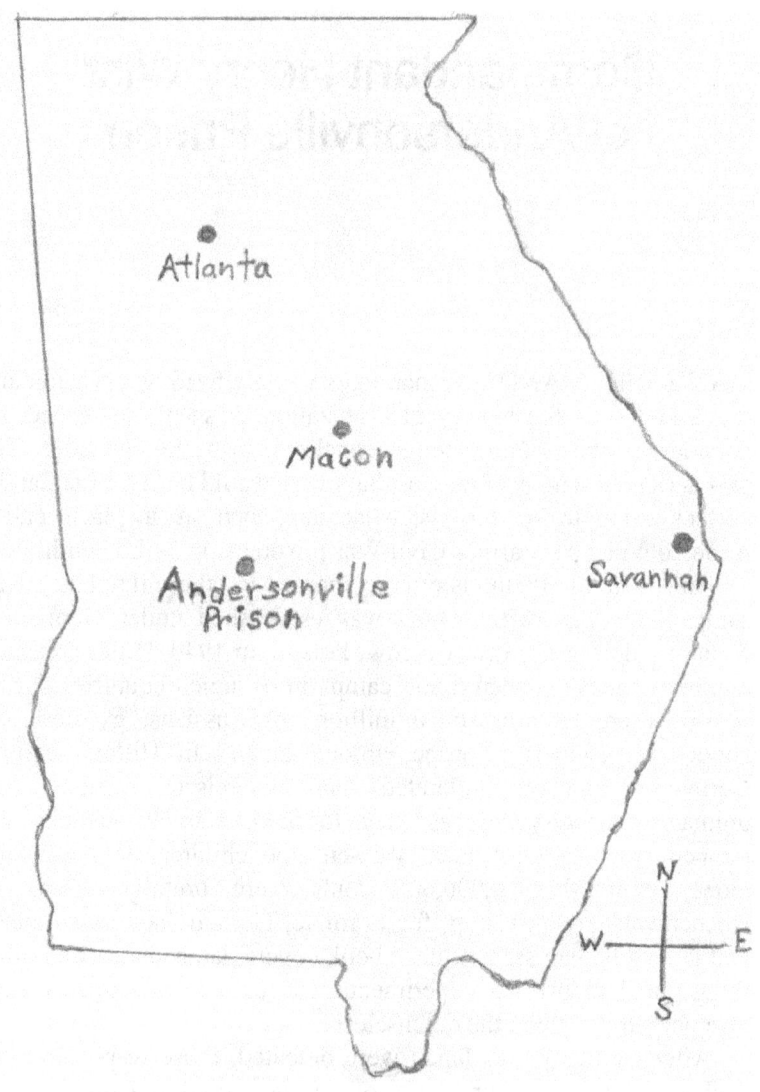

Location of Andersonville Prison in Georgia.

Commandant Henry Wirz

H. Winder serving as the Superintendent of Military Prisons for Alabama and Georgia was ordered to find a suitable place in Georgia for a prison camp; the site he selected was near the town of Americus. First known as Camp Sumter, Andersonville Prison was located on two hillsides with a stream of swampy water flowing through the middle of the 20-acre plot of land; on both sides of "Stockade Branch" were a few acres of scum-covered swamp. Later on, however, another spring called "Providence Spring" was discovered. The prison was also close to the railroad and had plenty of timber for building materials. The work began in January, 1864, with the stockade being constructed using slave labor from area farms who felled trees and dug trenches to set a solid, vertical wall of hand-hewn logs 20-feet high. Another wooden fence was also made about 12 to 16-feet high which ran within the main enclosure—any prisoner found to be within this area called the "deadline," was shot.

In March, 1864, Heinrich Hartmann Wirz was made commandant of Andersonville Prison, and so began his notorious Southern association with it. Wirz was born in Zurich, Switzerland, in 1822; he attended college there but went on to study medicine in Paris, France, and Berlin, Germany. He eventually migrated to America and settled in the "bluegrass state" of Kentucky where he practiced medicine. However, after the American Civil War erupted, Wirz joined the Confederacy but was eventually wounded at the Battle of Seven Pines, which occurred on May 31-June 1, 1862, a few miles east of Richmond, Virginia. Afterwards, he served in several prisons until he was assigned to Andersonville.

About how committed the South was to keeping its prisoners captive and out of the war, Confederate Brig. Gen. Winder reported on July 27, 1864, that: "The officer on duty and in charge of the battery of Florida artillery at the time will, upon receiving notice that the enemy has approached within seven miles of this Post, open fire upon the stockade…It is better that the last Federal be exterminated than be permitted to burn and pillage the property of local citizens, as they will do if allowed to make their escape from the prison."

The prison was guarded by rifle pits and artillery emplacements.

CURIOUS AND UNUSUAL CIVIL WAR STORIES

As many as 32,899 inmates were kept within the stockade, which was originally built for only 10,000. There were no barracks constructed to house the men—not even the sick or dying. The prisoners erected small lean-tos or huts called a "shebang" or dug holes in the ground to try and escape from the elements. The Stockade Branch was used for everything under the sun and became a stream of filth.

About the appalling conditions at Andersonville Prison, in *The Soldier's Story*, Warren Lee Goss wrote that: "There was a portion of the camp, forming a kind of swamp, on the north side of the branch, as it was termed by the rebels, which ran through the centre of the camp. This swamp was used as a sink by the prisoners, and was putrid with the corruption of human offal. The stench polluted and pervaded the whole atmosphere of the prison. When the prisoner was fortunate enough to get a breath of air outside the prison, it seemed like a new development of creation, so different was it from the poisonous vapors inhaled from the cesspool with which the prison air was reeking. During the day the sun drank up the most noxious of these vapors, but in the night the terrible miasma and stench pervaded the atmosphere to suffocation."

One account given about the prison told about seeing "forty or fifty men in a dying condition, who, with their little remaining strength, had dragged themselves to this place [the sink] for its convenience, and, unable to get back again, were exposed in the sun, often without food, until death relieved them of the burden of life. Frequently, on passing them, some were found reduced to idiocy, and many, unable to articulate, would stretch forth their wasted hands in piteous supplication for food and water, or point to their lips, their glazed eyes presenting that staring fixedness which immediately precedes death. On some the flesh would be dropping from their bones with scurvy; in others little of humanity remained in their wasted forms but skin drawn over bones. Nothing ever before seen in a civilized country could give one an adequate idea of the physical condition to which disease, starvation, and exposure reduced these men. It was strange that men should retain life so long as to be reduced to the skeleton condition of the great mass who died in prison."

Commandant Henry Wirz

In *A Prisoner of War in Virginia 1864-5*, George H. Putnam wrote that on August 5, 1864, Colonel D.T. Chandler, Inspector General, C.S.A., reporting about the medical horrors at Andersonville, said: "The acreage gives somewhat less than six feet square to each prisoner (that is, 2 feet by 3). Many (bodies) are carted out daily...whom the medical officers have not seen...The dead are hauled out daily by waggon loads and buried without coffins. Their hand in many instances being mutilated with an axe in removal of any finger rings they may have. It is impossible to state the number of sick, many dying whom the medical officers neither see nor hear of until the remains are brought out for burial."

There were many reports made about Andersonville Prison, Dr. Joseph Jones of the Medical Department, C.S.A., made these sad observations: "I visited two thousand sick within the stockade lying under some long sheds...At this time only one medical officer was in attendance, whereas at least twenty should have been employed...The sick lay upon bare boards or upon such ragged blankets as they possessed without...any bedding or even straw. The haggard distressed countenances of those miserable, complaining, dejected, living skeletons, crying for medical aid and food...and the ghastly corpses, with their glazed eyeballs staring up into vacant space, with flies swarming down their open and grinning mouths and all over their ragged clothes, infested with numerous lice, as they lay amongst the sick and dying, formed a picture of helpless, hopeless, misery which would be impossible to portray by words or by the brush. Millions of flies swarmed over everything and covered the faces of the sleeping patients and crawled down their open mouths and deposited their maggots in the gangrenous wounds of the living...Where hospital gangrene was prevailing it was impossible for any wound to escape contagion under these circumstances."

Captain Wirz was reported to have been domineering and abusive towards the inmates and many of them would have given their lives to put an end to him. The atrocities and death-toll at Andersonville has left the site, reportedly, haunted. Many people over the years have claimed to have seen ghosts and other things that go bump in the night—no doubt because of the horrible things that took place there which refuse to die.

The Andersonville Prison was open for a little more than a year—and, during that time, over 45,000 Union soldiers were incarcerated there—about 13,000 would die there as a result. Over the course of the Civil War (1861-1865), the Union captured and imprisoned about 220,000 souls; while the Confederates held about 210,000. In the end, about 26,000 Southerners died in Northern prison camps and about 22,000 Northerners perished in Southern camps; however, only one man paid the ultimate price for war-time atrocities: Henry Wirz.

After a trial that lasted for two months, Capt. Wirz was tried and convicted of his activities at Andersonville Prison during the Civil War and sentenced to undergo a neck-snapping death. On November 10, 1865, he went to the gallows and became the only Confederate official to be executed for war crimes. To the very end he claimed to be innocent, and that there was little he could have done to make things better at Andersonville because of overcrowding, lack of food and supplies, and other things. After the execution, his remains were taken to Washington, D.C., and buried at the Mount Olivet Cemetery. Some believed that he was convicted to divert attention away from the high death-count in Northern prisons.

There were some reports that witnesses committed perjury in order to condemn Wirz. In *Prisoners of War 1861-65: A Record of Personal Experiences, and a Study of the Conditions and Treatment of Prisoners on Both Sides During the War of the Rebellion*, Lieutenant Thomas Sturgis who served as a Union guard at Camp Morton near Indianapolis, Indiana in 1864, as well as being a prisoner himself at Libby Prison in Richmond, Virginia in 1865, delivered an address to the New York Commandry of Military Order of the Loyal Legion in 1911, where he told about the murders that had been perpetrated by Capt. Wirz while he was Commandant at Andersonville "with his own hand" in "cold blood."

Concerning the idea of Northern aggression towards Confederate prisoners, Lieut. Sturgis said: "It is worth noting that our Government did not swerve [according to Sturgis] from its humane policy for purposes of general retaliation."

Commandant Henry Wirz

In his address Lieut. Sturgis also said that "Mr. Lincoln was repeatedly urged by officials and officers of high rank to treat all rebel prisoners as our men were being treated. But this he steadily declined to do, saying that he would observe the wages of civilized warfare whatever our antagonists might do.

"And this was also the attitude of Congress."

Regardless of whether or not Wirz was the monster that he was reported to be, he was put to death. Wirz's hanging may have had a calming effect on a weary nation, but the ongoing questions over the matter of scapegoat or justice will no doubt continue to be a curious and unusual study of Civil War history. Nevertheless, on May 12, 1909, in the small hamlet of Andersonville, an impressive stone monument was erected by the United Daughters of the Confederacy. The monument stands less than a mile from the Andersonville National Cemetery and the historic site of the wartime prison.

CURIOUS AND UNUSUAL CIVIL WAR STORIES

Bibliography

Crain, Mary Beth, *Haunted U.S. Battlefields*, Globe Pequot Press, 2008.

Funk & Wagnalls New Encyclopedia, Funk & Wagnalls, Inc., New York, 1979.

Goss, Warren Lee, *The Soldier's Story*, Lee and Shepard, Boston, 1869.

Jackson, Rex T., *The Sultana Saga: The Titanic of the Mississippi*, Heritage Books, Inc., 2003.

Legrand, Jacques, *Chronicle of America*, Chronicle Publications, Mount Kisco, New York, 1989.

Moran, Mark, and Sceurman, Mark, *Weird Civil War: Your Travel Guide to the Ghostly Legends and Best-Kept Secrets of the American Civil War*, Sterling, New York, 2015.

Nesbitt, Mark, *Civil War Ghost Trails: Stories from America's Most Haunted Battlefields*, Stackpole Books, 2012.

Putnam, George H., *A Prisoner of War in Virginia 1864-5*, G.P. Putnam's Sons, The Knickerbocker Press, New York and London, 1914.

Ransom, John L., *John Ransom's Andersonville Diary*, Berkley Books, New York, 1994.

Sturgis, Thomas, *Prisoners of War 1861-65: A Record of Personal Experiences, and a Study of the Conditions and Treatment of Prisoners on Both Sides During the War of the Rebellion*, Report of an Address Delivered Before the N.Y. Commandry of the Military Order of Loyal Legion, 1911.

William "Bloody Bill" Anderson: The Demise of an Infamous Guerrilla

AMERICA'S Civil War is richly adorned with beneficial examples and priceless lessons—like the stain of slavery. A thorough study of it can offer crucial direction and comparisons to help as a guide for the present time as well as the future. This homegrown struggle that harvested and reaped the lives of about 633,000 participants produced a number of iconic figures both heroic and infamous; and records, stories and wild tales have followed on their heels. Some have entertained glory, respect and honor while others are remembered with disdain and disgrace—or, are misunderstood. One larger-than-life historical figure of the War Between the States whose work eventually led to his demise is the notorious William "Bloody Bill" Anderson.

William T. Anderson was born in Kentucky in 1839. He moved to Randolph County, Missouri when he was young and grew up near Huntsville. Anderson eventually moved to Kansas, where according to Hamp B. Watts who wrote a small book *The Babe of the Company* in 1913, said: "...his aged father had been made to dig his own grave, forced to stand over it, falling therein when shot...." It would only be the beginning of the rage and turmoil that stirred within this future Missouri guerrilla.

It was said of Bill Anderson that he was "broad of shoulder, hands and feet small, coal-black hair and eyes, sallow complexion,

Author's illustration of William "Bloody Bill" Anderson.

William "Bloody Bill" Anderson

high cheek bone, mouth broad with thin lips, Grecian nose, thin mustache and shaggy chin whiskers. His eyes in repose were dreamy, though restless—when aroused in battle, opening wide, piercing and gleaming as of fire. A sharp, cracked voice—when used in command, penetrating. Uncommunicative, almost to the point of reticence. He has been described as 'a tiger let loose.' In the midst of conflict with the enemy—true...."

As a possible explanation for some of Anderson's future actions was the imprisonment, for some reason, of his sisters Janie, Molly and Josephine who were taken by Federals and held in a rickety old building—as the story goes—at 1409 Grand Avenue in Kansas City, Mo. It was believed and reported that the building was undermined by a band of Kansas Jayhawkers and, on August 14, 1863, Josephine and Charity Kerr, a cousin of Cole, Jim, Bob and John Younger, were killed when the makeshift jailhouse collapsed upon them. To add more fuel to the fire, Union General Thomas Ewing's Military Order No. 11, that left several western Missouri counties a burnt wasteland, further fired-up pro-Southern guerrillas like Anderson; at the time there were many things going on that split and divided the nation. To explain the place that Anderson had arrived at, Watts wrote: "Revenge and revenge alone, permeated and took possession of every fiber of Bill Anderson's body. He had come to the place at last where an eye had to be rendered for an eye and a tooth for a tooth. He found nor asked no quarter—he gave none...."

Joining William Clarke Quantrill, Bill Anderson became a valued member as one of his lieutenants and participated on August 21—a few days after the collapse of the building in Kansas City, in the sacking and massacre of Lawrence, Kansas, where about 150 men and boys were attacked and killed—Lawrence was the hometown of the hated Senator Jim Lane who somehow managed to escape the raid—had he been captured, Quantrill had plans to take him back to Missouri to face the justice of the noose or to be burned at the stake. As a result of the atrocities perpetrated upon Lawrence, there was national outrage; had the nation showed more sympathy and compassion for the girls that had been killed or injured in the collapsed jail, possibly the rage of retribution might have been lessoned, to some degree, and history altered.

CURIOUS AND UNUSUAL CIVIL WAR STORIES

According to William Esley Connelley in his book *Quantrill and the Border War*: "Every guerrilla carried two revolvers, most of them carried four, and many carried six, some even eight. They could fire from a revolver in each hand at the same time. The aim was never by sighting along the pistol-barrel, but by intuition, judgment. The pistol was brought to the mark and fired instantly, apparently without a care, at random. But the ball rarely missed the mark—the center. Many a guerrilla could hit a mark to both the right and the left with shots fired at the same instant from each hand.

"No more terrifying object ever came down a street than a mounted guerrilla wild for blood...his long unkempt hair flying wildly beyond the brim of his broad hat, and firing both to the right and left with deadly accuracy. When a town was filled with such men bent on death, terror ensued, reason and judgment fled, and hell yawned."

By 1864, however, Anderson's men, recruited from the Missouri counties of Clay, Jackson, LaFayette, Cass and Johnson, were making a name for themselves, for the most part, north of the Missouri River while Quantrill worked south of it—both of these guerrilla bands were commissioned by the Confederacy.

Union cavalry scouting parties made frequent expeditions and maneuvers along roads and highways with "posts at Columbia, Fayette, and Boonville, numbering in each foray from two to three hundred troopers." The Federals kept the guerrillas "alert and moving" and often exchanged gunfire, but on many occasions "guerrillas always found safe retreat in the brush."

Most of Anderson's barbarous work went on in the counties of Clay, Ray, Carroll, Chariton and Fayette. On September 27, 1864, "Bloody Bill" Anderson, as he had become known, took about thirty men of his force into Centralia, Mo. to terrorize its inhabitants. For several hours they kept the town entertained with fear, burning down the train depot and other things. A stagecoach from Columbia arrived which provided the guerrilla fighters a few more victims to torment. Eventually a train from St. Charles, the North Missouri Railroad, arrived and its passengers were forced from the train, including twenty-five Federals; twenty-four of the soldiers were

William "Bloody Bill" Anderson

mercilessly killed and the train was burned. Watts recalled that "The murder of the unarmed and defenseless men taken from the train was a relentless act of war, without palliation or excuse, heartless in its nature, savage in its execution, their only crime being the uniform they wore."

Afterwards, Union Major A.V.E. Johnston and his 39th Missouri Infantry arrived in Centralia in late afternoon and found Anderson's handiwork. According to Wiley Britton in his book *The Civil War on the Border*, "The blood was still oozing from the wounds of the murdered soldiers, and in some instances their clothes was still burning and their bodies scorched to a crisp."

It was reported that Major Johnston "bragged how he was going to extinguish" their whole force and, "showing his black flag" he announced that he would "take no prisoners" and would hunt them down and return with Captain George Todd's and Bloody Bill Anderson's "heads tied to a pole." This may have been music-to-their-ears for the people of Centralia.

About one-and-a-half miles southeast of Centralia was a patch of timber that was believed to be inhabited by Anderson's murderous band of guerrillas. Major Johnston marched his Federals towards the woods to carry out his planned extermination, all the while unaware that an ambush of well-armed, wild-for-blood bushwhackers were anxiously waiting for them with "bells on".

When Johnston's force came within two or three hundred yards of the timber, about 400 guerrillas charged out, "yelling like fiends, with their bridle-reins in their teeth and a revolver in each hand." A few men in the rear escaped, but over one hundred Federal soldiers were massacred. Frank and Jesse James were in Anderson's force, and Jesse is credited to having killed Major Johnston.

A few days later on October 14, 1864, Bloody Bill and about 60 armed horsemen descended upon Danville, Mo. and shot citizens and burnt down buildings but spared the Female Academy on the edge of town. Afterwards, they headed east on the "Booneslick Trail" until they arrived at the Baker Plantation House—the two-story Baker House was built of bricks by slave labor in 1850 by Sylvester Marion Baker. The raiders demanded that Mr. Baker come out of the house and face them, however, Mrs. Frances Baker, with child in arm, opened the door instead and told them that her husband was not at home and that they could search the house. The guerrillas made their search; rode their horses through the foyer; stole what they wanted; and set fire to the Baker's bed and parlor room before leaving—somehow, the fire was eventually put out.

Not long after this event, on October 27, 1864, near Richmond, Mo., in Ray County, William "Bloody Bill" Anderson was killed. In an official report to Union Brigadier General James Craig which ran in Arkansas' Fort Smith *New Era* on November 26, 1864, it offered the news breaking details to its readers about the demise of the infamous Missouri guerrilla: "We have the honor to report the result of our expedition yesterday against the notorious bushwhacker Bill Anderson and his force, near Albany [present-day Orrick, Missouri] in the southwest corner of this country [county]."

Union Major Samuel P. Cox of Gallatin, Mo., and a force of about three hundred men armed with muzzleloaders and revolvers had marched there from nearby Richmond. About three miles later, Major Cox met up with a woman on horseback who said that she knew where Anderson's camp was located and that he had about two or three hundred well-armed men bivouacking with him.

The official report in the *New Era* went on to say that: "Learning his whereabouts, we…made a forced march, coming in contact with their pickets about a mile this side of Albany [Orrick]. We drove them in through the town and into the woods beyond, when we dismounted our men and threw an infantry force into the forest, sending a cavalry advance who soon engaged the enemy and fell back, when Anderson and his fiendish gang, about 300 strong, raised the Indian yell and came in full speed upon our lines, shooting and yelling all the while. Our lines held their position without break.

"Anderson and one of his men, supposed to have been Captain Rains, son of General [James] Rains, charged through our lines. In this charge Anderson was killed, falling some fifty steps in our rear, having received two balls in the side of the head. Rains made his escape, and their forces retreated at full speed, being completely routed….

"We captured on Anderson's body private papers and others

William "Bloody Bill" Anderson

from Gen. [Sterling] Price, that identify him beyond a doubt.

"Upon the body of the brigand Anderson was found $300 in gold, $150 in Treasury notes, six revolvers, and several orders from General Price...."

Orders from Confederate Major General Sterling Price authorized Anderson to work at destroying the North Missouri Railroad and to torch and destroy property. Also among Bloody Bill's papers was a commission signed by Confederate President Jefferson Davis which elevated the guerrilla chieftain to the rank of colonel.

Jesse James was numbered in Col. Anderson's force that day, but managed to escape. One author who wrote about such men like Quantrill, George Todd, and Bill Anderson, offered this observation: "All three died as they had lived with 'boots on,' worshiped by a few, loved by many, and abhorred of half the Nation."

Afterwards, the lifeless corpse of William T. Anderson was taken to the Richmond, Mo. courthouse and kept under guard and buried the next day at a cemetery near town. No more did the "bugles blast, the saber's glare, the cannon's roar, the whirring bursting shell, the rattle of musketry, the whizzing Minnie ball, the sharp, quick crack of deadly revolver—the groan of wounded and dying comrades" plague him or visit him. For such men the sorrows "that can never be obliterated" ended once and for all in the spring of 1865. The years of that bloody American struggle brought forth a number of colorful, memorable characters who did what they did for many and various reasons—history, however, remembers them as legends, heroes, icons—and, in some cases, connoisseurs of slaughter.

Bibliography

Barton, O.S., *Three Years with Quantrell: A True Story Told by His Scout John McCorkle*, Armstrong Herald Print, 1914.

Britton, Wiley, *The Civil War on the Border*, Vol. 2, G.P. Putnam's Sons, The Knickerbocker Press, New York and London, 1899.

Connelley, William Esley, *Quantrill and the Border Wars*, The Torch Press, Cedar Rapids, Iowa, 1910.

Crowson, Noel and Mary Ann, *Baker Plantation House*, Blue & Gray Magazine, Vol. 18, 2000.

Fort Smith *New Era*, November 26, 1864.

Jackson, Rex T., *Notable Persons and Places in Missouri's History*, The Ozarks Reader, Neosho, Missouri, 2006.

Settle, Jr., William A., *Jesse James Was His Name*, University of Missouri Press, 1966.

Triplett, Frank, *The Life, Times, and Treacherous Death of Jesse James*, St. Louis, J.H. Chambers and Company, 1882.

Battle of Carthage: First Significant Land Battle of the Civil War

LONG BEFORE adventurous road warriors traveling Route 66 cruised through Carthage, Missouri headed east and west, it was a victim of the North and South during the American Civil War. More than two weeks before the 1st Battle of Manassas (or Bull Run) and the Union's presumptuous battle cry of "On to Richmond," the first significant land battle of the War Between the States had already occurred at Carthage, Mo. This historical distinction, in the shadows of the great Eastern battles—and nearly overlooked by many heralded historians and Civil War buffs, was highlighted and featured, to some degree, in the New York *Times* when it cried and informed its readership about the seriousness of this first significant Civil War battle in the West.

On April 12, 1861, the first shots of the Civil War were fired by a circle of five Confederate batteries commanded by Confederate General Pierre Gustave T. Beauregard, under orders of Confederate President Jefferson Davis, on Fort Sumter off Charleston Harbor, South Carolina. However, this seacoast bombardment was delivered about two months before the first serious land battle at Carthage—Fort Sumter earning its own Civil War distinction.

As for Missouri and the upcoming engagement at Carthage, the capture of the St. Louis Arsenal at St. Louis on May 10, 1861, by Union Captain Nathaniel Lyon, prompted the pro-Southerners under

CURIOUS AND UNUSUAL CIVIL WAR STORIES

Governor Claiborne Fox Jackson and ex-governor Major General Sterling Price to regroup and issue a warning to the federal government about launching an invasion. Nevertheless, a newly promoted Brigadier General Lyon pursued them and caught up with them on June 17, 1861, at Boonville, Mo. where an attack was made. The skirmish at Boonville became the earliest land battle of the Civil War, and left Gov. Jackson and Maj. Gen. Price in full retreat southward towards Carthage and closer to the safety of Confederate Arkansas.

By July 3, 1861, Gov. Jackson and Gen. Price's Missouri State Guards totaling in all about 6,000 men, with only 4,000 armed and seven pieces of artillery, were at Lamar, Mo. and on their way south towards Carthage; however, Gen. Price had taken his leave temporarily and had left Brigadier General James S. Rains in command of his troops. In the *Official Records of the Union and Confederate Armies*, Confederate Brig. Gen. James S. Rains reported that the Southern "force consisted of the First Brigade, commanded by Colonel Weightman, of the First Cavalry. This brigade was composed of Capt. Hiram [Miller] Bledsoe's company of artillery (three pieces—one 12-pounder and two 6-pounders), 40 men, and Captain McKinney's detachment of infantry, 16 men, commanded by Lieutenant-Colonel Rosser, of the First Infantry; Colonel Graves' independent regiment infantry, 271 men; Colonel Hurst's Third Regiment Infantry, 521 men, and Lieutenant-Colonel O'Kane's battalion of infantry, 350 men, being in all 1,204 strong.

"The cavalry brought on the field consisted of Companies A and B and part of H of the Third Cavalry, 115 men, commanded by Colonel Peyton, to whom was attached the companies of Captain Stone and Owens. The First Battalion of the Independent Cavalry, 250 men, commanded by Colonel McCown; Lieutenant-Colonel Baughn's [Vaughn] battalion of the Fourth Cavalry, 200 men, and Capt. Joseph O. Shelby's company of Rangers, 43 men, making a total of 1,812 men. The remaining portion of my command, being unarmed, was used to present the appearance of a reserve corps and baggage guard."

Also in the area of Carthage was the German-American Colonel Franz Sigel and his Union force of about 1,100 camped southeast of

Battle of Carthage

A reenactment photo taken at Carthage, Mo.

town "behind the Spring River." Col. Sigel reported that his command consisted of "Nine companies of the Third Regiment, with a total effective strength of 550 men; seven companies of the Fifth Regiment, numbering 400 men; two batteries of artillery, four pieces each."

On the morning of July 5, 1861, Sigel's force crossed Dry Fork Creek about 6 miles north of Carthage and pressed on about 3 miles more where they "found the enemy in line of battle on an elevated ground, gradually rising from the creek, and about one and a half miles distant."

In *The First Year of the War in Missouri* in *Battles and Leaders of the Civil War*, Confederate Colonel Thomas L. Snead recalled the scene that morning, and wrote: "Halting until the 5th in order to rest and organize, they pushed on that morning toward Carthage, having heard that a Federal force had occupied that place, which lay in their line of retreat. They had marched but a few miles when, as they were passing through the open prairie, they descried, some three miles away, on the declivity of a hill over which they had themselves to pass, a long line of soldiers with glistening bayonets and bright guns."

Sigel's Federals were outnumbered about four to one but he did not hesitate to bring "his troops into action to the sound of music and in perfect step."

Brig. Gen. Rains sent out orders for Capt. Shelby to check on the Union advance and for Col. Weightman to "deploy the brigade in order of battle on the ridge of prairie overlooking the enemy."

After a few spherical-case shots were launched in Shelby's direction, his company was sent to the far right to look for a way to cross the creek and possibly flank the enemy. According to Brig. Gen. Rains, Shelby's movement was "conducted in the face of both armies" and "executed with a precision worthy of the parade ground."

Around 9:00 a.m. Col. Sigel broke the morning silence with the thunderous booming and din of his big guns, "throwing grape, canister, shell, and round shot" to start the 1st Battle of Carthage; also known as the Battle of Dry Fork by the Confederates. Soon after, the Southern batteries and the crack, crack, crackle of

Battle of Carthage

musketry was returned upon the Union and the air was full and saturated with black powder battle-smoke and missiles of every description.

One of Confederate Capt. Bledsoe's cannons deployed that day was the well-known "Old Sacramento," an ancient Mexican fieldpiece that was confiscated during the Mexican War (1846-1848) when Bledsoe was under Alexander W. Doniphan's "famous cavalry, whose prodigious marches and dashing combats adorn the brightest pages of American history." It was reported that the old gun had been cast and smelted from Chihuahua brass church bells, with silver added to the mix. After it was brought back to Missouri by Doniphan, along with other war relics, it was put to use for Fourth of July celebrations and salutes in Lexington and other Missouri River towns.

When the Civil War broke out Bledsoe naturally drafted the ancient cannon into the Confederacy. The 9-pound gun was bored out and made into a 12-pound howitzer. "The chase was turned off smooth, thus reducing the thickness of the metal, which gave the piece a peculiar sound when fired...." During the Battle of Carthage Bledsoe reportedly fired "trace chains, old scrap-iron, and smooth pebbles" out of the muzzle of the old ox-drawn cannon. The cannon's reputation at Carthage apparently carried over to other battles and skirmishes and, at the Battle of Wilson's Creek which occurred near Springfield, Mo. on August 10, 1861, Old Sacramento filled the battleground with fear and terror. One German-Federal trooper reportedly asked: "Vare iss der man mit der ogs gannon?" But after Bledsoe's powerful ancient, customized gun was fired, the German quickly shouted: "Mine Gott in Himmel!"

Bledsoe's batteries would become famous during the war and many attempts were made to capture or silence his deadly handiwork. Many times the ammunition was improvised. "Cartridge-bags were sewed, canisters cut...In lieu of grape-shot, canisters were filled with iron slugs, trace-chains—anything a country blacksmith shop could supply. This was called 'scrap-shot.' Most of the shells and solid shot were spoils of battle, nearly every engagement furnished a supply for the next."

After Sigel's "well-directed fire" was concentrated upon the

Confederate advance, the battle-tide began to change and fearing that his force could be outflanked, and also suffering from the rapid-fire and the accuracy and precision of Bledsoe's booming batteries, he decided it was high-time to hightail it back across the creek.

Gov. Jackson's state troops were impressed by the "precise movements and soldierly bearing" of Sigel's Federals; however, the overwhelming numbers of the Southerners soon put the Union to flight—forcing them to retreat from the Dry Fork Creek area. They retired to Buck Creek where the action resumed for a spell, but were forced to continue their retreat to Spring River where Brig. Gen. Rains' cavalry and General Slack's division under Colonel Rives "endeavored to out flank them on the right, but the retreat" was too fast for them.

As the Confederates neared Spring River they attempted to stop the Federals from crossing, but Sigel's guns were too much for them and they were forced to cross the river further upstream and, eventually, the battle was taken into the streets of Carthage. Overtime, Sigel's Federals were in full retreat "carrying off almost everything that he had brought with him. But he didn't stop running till he had made forty miles." The Confederates savored their victory and bivouacked that night without opposition in Carthage. Col. Franz Sigel's plan of throwing himself "in front of the Governor, hoping either to defeat him or hold him in check" until Captain Nathaniel Lyon could arrive with reinforcements "and destroy him," had failed.

About the work of Gov. Jackson's State Guards, in *Battles and Biographies of Missourians* by W.L. Webb, Col. Sigel was reported to have said: "Great God! Was the like ever seen? Raw recruits, unacquainted with war, standing their ground like veterans, hurling defiance at every discharge of the batteries against them, and cheering their own batteries whenever discharged. Such material properly worked up would make the best troops in the world."

The Federal retreat from Carthage on the Sarcoxie Road was covered by artillery and infantry, which took their positions on the "heights behind Carthage...." Sigel reported that their losses were light, but estimated that the Confederate losses were about 350 to 400 killed or wounded. He summed it up in his report by saying that

Battle of Carthage

they were "more than once menaced in flank and rear by large forces of cavalry, and attacked in front by an overwhelming force, they stood like veterans, and defended one position after the other without one man leaving the ranks."

One remaining icon of the Civil War era in present-day Carthage is the Kendrick Place, located one mile north of town. The antebellum house was utilized as a headquarters during the Civil War. It was built by Jacob and William Rankin, who were cattlemen from Ohio, using slave labor beginning in 1849 and completed in 1853. The two-story Kendrick House was built of clay bricks taken from nearby Spring River and fired on the property. The 600-acre plantation-style farm was worked by 13 slaves who were on record at the Jasper County Courthouse for tax purposes.

In 1856 the Kendrick Place was sold to Thomas Dawson, a son-in-law of Sinnet Rankin, for $7,000. One again, the home and 540-acres was sold in 1860 to William B. Kendrick, and it remained with the Kendrick family and their descendants for over 130 years.

Maj. Gen. Sterling Price, who had been in Arkansas inviting Confederate General Ben McCulloch to "march into Missouri," rejoined the troops who were in "ecstasy at seeing their great captain." Price "assumed command and marched to Cowskin Prairie" in McDonald County, Mo. to train and reorganize, "where there was grass for the horses and lean beef for the men."

About a month later Price's army would face Gen. Lyon and Col. Sigel at the Battle of Wilson's Creek; it was also an early victory for the South—but short-lived, the tide would change in favor of the Union at the Battle of Pea Ridge (or Elkhorn Tavern) in northern Arkansas in March 1862.

The Battle of Carthage pales in comparison to great Civil War struggles like Shiloh, Antietam and Gettysburg, but it holds its own special place in significance as being the first serious land battle waged during the War Between the States—a distinction that will forever highlight it in American history.

Kendrick House located one mile north of Carthage, Mo.

Battle of Carthage

Bibliography

Jackson, Rex T., *The Battle of Carthage: The First Significant Land Battle of the Civil War*, Vol. 13, No. 4, Route 66 Magazine, 2006; *Historic Carthage: Nostalgic Road to the Civil War*, Vol. 1, No. 1, The Ozarks Reader Magazine, Neosho, Missouri, 2004.

Official Records of the Union and Confederate Armies, Washington: Government Printing Office, 1881.

Snead, Thomas L., *The First Year of the War in Missouri*, Vol. 1, Battles and Leaders of the Civil War, The Century Company, 1887.

Webb, W. L., *Battles and Biographies of Missourians* or the *Civil War Period of Our State*, Hudson-Kimberly Publishing Company, Kansas City, Missouri, 1900.

Fort Henry Surrenders to the Navy

THOUGH MANY noteworthy and unusual things occurred over the course of the American Civil War, one significant conflict, the Battle of Fort Henry, earned at least two distinctions: first, it was the only formal surrender of a land-based fortification solely to the Navy, which would not happen again throughout the war; and second, the engagement was the first successful use and victory of ironclad gunboats. (The famous duel between the iron-plated *Merrimac* (or *Virginia*) and the *Monitor* took place about a month after the Battle of Fort Henry and ended as a draw.)

On the south side of St. Louis, Missouri, at a place known as the Carondelet Shipyard is where James B. Eads, the old daring civil engineer, built a number of ironclads for the Union during the Civil War; others were also constructed across the Mississippi River and south at Mound City, Illinois. The first seven of Eads' riverine iron "turtles" were called: *Carondelet*, *St. Louis*, *Louisville*, and *Pittsburgh*, which were built in the Carondelet Shipyard; the other southern Illinois boats were dubbed *Mound City*, *Cincinnati*, and the *Cairo*.

In early February, 1862, Flag Officer Andrew H. Foote in a temporary flagship *Cincinnati*, commanded by R.N. Stemble, were on their way on the Tennessee River headed towards Fort Henry fishing up dangerous Confederate torpedoes (mines) out of the river as they crept along. The explosive mines were about 5½ feet long and 1 feet in diameter. They had an iron lever 3½ feet long with

Author's illustration of one of James B. Eads' city-class gunboats.

Fort Henry Surrenders to the Navy

prongs that could catch the bottom of passing boats. The lever acted as a trigger mechanism to set off the cap and detonate 70 pounds of powder. The device was anchored to the river-bottom to keep it in place. The Union flotilla, besides the flagship *Cincinnati*, consisted of the *Carondelet*, under Commander Henry Walke, the *St. Louis*, under Lieutenant Commander L. Spaulding, the *Essex*, under Commander W.D. Porter, and three timberclads, the *Tyler*, under Lieutenant Commander William Gwin, the *Lexington*, under Lieutenant Commander J.W. Shirk, and finally, the *Conestoga*, commanded by Lieutenant Commander S.L. Phelps.

On February 6, 1862, the Union naval force was nearing Fort Henry. The fort was located on the east bank of the river in a marshy area. On the opposite side of the river was Fort Heiman, an unfinished fortification occupied by about 1,100 Southerners—the Twenty-seventh Alabama Regiment, Fifteenth Arkansas, a couple of companies of Alabama cavalry, an unorganized company of about 40 Kentucky cavalry, and a section of a light battery; however, the majority of the Fort Heiman force was moved across the river to Fort Henry the day before in order to "beef" up its defenses—Fort Henry now had about 3,200 troops to defend itself against the powerful flotilla.

When the Union gunboats came within about 1,700 yards of the fort, the flagship *Cincinnati* fired off the first round to signal the fleet to attack. Suddenly the gunboats belched-out a din of heavy shot and shell from their fifty-four big guns. In return, Fort Henry's eleven artillery pieces, commanded by Brigadier General Lloyd Tilghman, poured out its own death-dealing cannonade as well into the Tennessee currents and into the fleet of ironclads.

The powerful iron-plated turtles kept up a rapid fire and pounded the fort's earthworks and sandbags. A number of guns were hit and buildings were set ablaze within Fort Henry. The arching shot and shell threw debris into the Tennessee air, while the Confederates responded with deadly accuracy upon the Union riverboats. The action eventually closed to within about 600 yards of the two combatants—one in the water and the other on the land. About twenty minutes before the battle ended, the *Essex* took a shot in its boilers which exploded and scalded 29 officers and men.

In less than two hours duration, Gen. Tilghman, in making an assessment of the number of guns still in working order and the condition of the fort, made the command decision to surrender. The rebel flag was lowered and Gen. Tilghman and a few of his staff took a small boat out to the *Cincinnati* and officially surrendered to Flag Officer Foote, but not before he had dispatched most of his troops to nearby Fort Donelson; only about 70 men were left at Fort Henry to be taken prisoner.

The celebration and cheer of the Union victory quickly disappeared when they, upon reaching the fort, saw that "On every side the blood of the dead and wounded was intermingled with the earth and their implements of war."

The plan, which originally included the ground forces of General Ulysses S. Grant, to join in on the naval attack at Fort Henry, never materialized because of the muddy roads and high water that hampered their timely arrival. As a result, the surrender of Fort Henry was accepted by Flag Officer Foote and his inland navy.

Confederate General A. Sidney Johnson, commanding the Western Department, reported on February 8, 1862, two days after the battle, that: "The slight resistance at Fort Henry indicates that the best open earth-works are not reliable to meet successfully a vigorous attack of ironclad gunboats...."

Fort Donelson on the Cumberland River was the next in line to fall to the Union. On February 14, 1862, eight days after Fort Henry, ironclads would again prove to be a formidable military adversary at Fort Donelson; however, a few days later on March 9, 1862, at Hampton Roads, Virginia, the *Merrimac* and the *Monitor* would continue the new iron tradition and become a media sensation which drowned out the clash and noise of the earlier ironclad bombardments of "Henry" and "Donelson."

Bibliography

Coombe, Jack D., *Thunder Along the Mississippi: The River Battles That Split the Confederacy*, Bantam Books, 1996.

Dorsey, Florence, *Road to the Sea: The Story of James B. Eads and the Mississippi River*, Rinehart & Company, Inc., New York and Toronto, 1947.

Jackson, Rex T., *James B. Eads: The Civil War Ironclads and His Mississippi*, Heritage Books, Inc., Bowie, Maryland, 2004.

Official Records of the Union and Confederate Armies, Washington: Government Printing Office, 1881.

Taylor, Jesse, *The Defense of Fort Henry*, Battles and Leaders of the Civil War, The Century Company, 1887.

Walke, Henry, *The Gun-Boats At Belmont and Fort Henry*, Battles and Leaders of the Civil War, The Century Company, 1887.

Women in the Civil War: Fighting the Stigma of Weakness

THE CALL of glory, patriotism, religion, honor, adventure, service, and a host of other things guided many to volunteer to fight during the American Civil War—for both the North and the South. Eventually, after the flood of news from the battle-lines kept pouring in, much of the earlier enthusiasm started to dry up. As a result, a draft was established to maintain the ranks of "Billy Yank" and "Johnny Reb." However, it wasn't only men that marched off to war and the beat of the drum, some women felt the need to participate also. Without the approval and knowledge of their governments, several hundred women dawned manly apparel and, disguised as their male counterparts, served during those bloody years of the War of the Rebellion (1861-1865) in the Union and Confederate armies.

On the home-front, women made sacrifices and contributions far above the call of duty, working in needed and important manufacturing, making uniforms and socks, working with military and medical supplies, on the farm, caring for family and friends and many other things. One old, popular song went like this:

> "Just take your gun and go,
> For Ruth can drive the oxen, John,
> And I can use the hoe."

Civil War reenactment photo taken at Bentonville, Arkansas.

Women in the Civil War

The greatest and most compassionate service rendered by women during the Civil War was in the field of nursing. The male-dominated medical profession, at the time, held women in low esteem when it came to their abilities; however, as time passed on and women continued to persevere in the face of this unwarranted, chauvinist, prejudice view, the stigma began to suffer and die—women began to fill the nursing ranks. The work they championed, though, was beset by the horrors of war where they endured carnage and unspeakable sights and sounds associated with battlefields and hospitals—yet, they did it without reservation for the sake of the sick and injured.

In *With Sabre and Scalpel*, John Allan Wyeth offers an eyewitness example of war's reality: "...I came to one of the field hospitals where the surgeons were busy with the wounded, stretched out on their blankets under the trees. One poor fellow was walking up and down holding the freshly amputated stump of his forearm with the remaining hand. His jaws were firmly set, and his face wore the hard, fixed expression of pain...Some fragments of arms and legs lying around completed the gruesome picture."

An excellent example of this is Mary Whitney Phelps who was born in Portland, Maine, in 1813. Mary eventually married John Smith Phelps in 1837; John served as governor of Missouri from 1877 to 1881. The Phelps' made their home in Springfield, Mo., and had a sizable acreage.

After the Battle of Wilson's Creek (or Oak Hills) on August 10, 1861, which occurred near Springfield, Mary supervised the burials of many of the victims of the conflict and did much of the grievous work herself. The body of Union General Nathanial Lyon, who was the first general to fall during the Civil War, was brought to the Phelps farm for safekeeping. Since Mr. Phelps was not present at the time, Mary took charge and assumed responsibility. Knowing that the general's remains would eventually be sent for, the body was placed temporarily in an outdoor root-cellar (some refer to them as "caves") which was used as an icehouse in the summer months and as an "apple-hole" during wintertime; straw was also incorporated to further preserve Lyon's body. Word traveled fast that the general was at rest in the makeshift morgue and, since the

CURIOUS AND UNUSUAL CIVIL WAR STORIES

Confederates had won the Battle of Wilson's Creek and were still encamped in the area, a number of citizens and soldiers turned out for a look-see.

One Confederate soldier took the liberty of offering his opinion of Gen. Lyon to Mrs. Phelps, saying: "There is quite a contrast betwixt the resting place of old Lyon's body and his soul, isn't there, Madame? The one is in an ice-house; the other in hell!" Other Southerners also made vile comments which prompted Mary, fearing the safety of the body, to appeal to Confederate Major General Sterling Price (himself a former governor of Missouri) to lend a hand in the care of the fallen Union general. Gen. Price dispatched volunteers of General Monroe M. Parsons' division for the task; the body of Gen. Lyon was removed from the cave and buried in the Phelps' garden spot. However, the trouble continued at the expense of the fallen general as the workers reportedly tromped and stomped upon the shallow grave in disrespect. One Irish soldier confessed in utter delight: "Be jabers, we shtomped him good."

Eventually, a four-mule-train ambulance arrived toting a 300-pound metallic coffin and returned the body of Gen. Lyon to his birthplace in Eastford, Connecticut, where he finally received an honorable burial.

Mary was also on hand after the bloody Battle of Pea Ridge fought near Pea Ridge, Arkansas, on March 7-8, 1862, as a quartermaster commissary, and as a nurse caring for the wounded; she went on to receive an award from the United States Congress for her brave, well-deserved contributions during the war.

Mary didn't stop there, she established a postwar orphanage in Springfield, helped to get Confederate remains buried, and in 1870 became vice president of the National Women's Suffrage Association—helping to further the rights of women; at the time, even voting for women was, for the most part, too liberal for Christian-conservative America. Mary died on January 15, 1878, and was buried with her husband at the Hazelwood Cemetery in Springfield.

Another woman that participated in the Civil War struggle was Rose O'Neal Greenhow, a well-to-do socialite and widower who was Southern-born and became a spy for the Confederacy.

Women in the Civil War

Greenhow relocated to Washington, D.C., in order to be able to eavesdrop on the "Yanks." The plan went well and in July 1861, she got "wind" that the Federals under Brigadier General Irvin McDowell were headed for Manassas, Virginia. Greenhow wasted no time alerting Confederate General Pierre Gustave T. Beauregard that the Northern invasion was on the way.

Greenhow continued her espionage collecting pertinent information for the South until she was arrested on August 23, 1861, and incarcerated at the Old Capital Prison; she was eventually released on May 31, 1862, and deported to Richmond, Virginia, where she was heralded as a heroine of the Southern cause. Before long, President Jefferson Davis of the Confederate States of America called upon her once again, this time to travel to Europe to raise monetary aid and support for the South. However, on the return voyage, just off the coast of North Carolina, a Union warship in search of blockade runners ran her ship aground. Greenhow attempted desperately to reach shore but, because she was smuggling gold that was sewn into the lining of her clothes, the weight of the precious metal became too much of a burden and she drowned.

Thousands of women went to the battle-front to serve as nurses, where they endured the many hardships of camp-life and the horrific realities of war to offer mercy and healing to their countrymen. A perfect example of this can be found in Clarissa Harlowe Barton— better known as Clara Barton. Clara was born in 1821 in Oxford, Massachusetts, and was educated, for the most part, by her two brothers and two sisters. In her earlier years, Clara spent time teaching and helped to establish a number of free schools in New Jersey. She found employment in 1854 at the Patent Office in Washington, D.C., but by 1861 at the outbreak of the Civil War she loaded up a wagon with medical supplies and other necessary items and headed for the bloody battlegrounds—and, at her own peril, began to treat the wounded. Clara reportedly utilized a pocket-knife to dig bullets out of soldiers; she also dressed wounds with bandages or with whatever was available, like cornhusks. She was sometimes referred to as the "angel of the battlefield."

In the years that followed the end of the war, Barton helped to do

Author's sketch of Sarah Emma Edmonds.

Women in the Civil War

systematic searches for soldiers that were missing. As a result of the ongoing work, the American Red Cross Society was founded; she served as the first president of the Red Cross from its beginning in 1881 to 1904. As a philanthropist, Barton did many charitable things during her lifetime and passed away in 1912.

Yet another female heroine of the War Between the States was Sarah Emma Edmonds, whose unusual tactic of masquerading as a man in order to enlist and fight in the war, went on to make her a legend and icon to many Civil War history-buffs. Sarah was born in Nova Scotia, Canada in 1841, her father, who desperately wanted a son, never failed to make his disappointment known to his daughter. It's not hard to imagine or to understand that on April 25, 1861, Sarah Emma Edmonds, under the assumed identity of Frank Thompson, headed for the United States of America, lopped-off her hair, dressed up as a man, and enlisted in the 2nd Michigan Volunteer Infantry to fight in the Union army—not something the governments of either the North or the South sought or desired. Edmonds would join an elite, historical group of women (about 300 for the North and about 150 for the South) that fought secretly beside men. In most cases, their feminine identity was never disclosed except by capture or death.

Edmonds was assigned to nursing and served in a number of major battles, but she wanted more, to participate in the dangerous game of espionage for the Union. With the aid of silver nitrate to help darken her color of skin—since whites, for some reason, considered blacks more suitable for slaves—Edmonds took on the persona of a slave behind enemy lines and reported back useful intelligence to the Union. Like a "chameleon" she utilized different disguises to fool the Confederates and gather information vital to war efforts.

Eventually, however, Edmonds contracted malaria, and instead of seeking care in an army hospital, went to a private facility in Illinois with the hopes of her true gender not being discovered. In her absence, however, she learned that Frank Thompson was being sought for military desertion, so she decided to go to Washington and find employment as a nurse being herself.

After the war ended in 1865, Edmonds returned to her Canadian homeland and married a longtime acquaintance, Linus Seelye, and had three children. Her wartime service—in spite of her father's opinion of having a daughter instead of a son—was eventually acknowledged by the government and she received an honorable military discharge and pension. Sarah Edmonds died at the age of 57 in 1898. Her service is a testament to the bravery, strength, and character of all women burdened with the unfair stigma of weakness that has been placed and piled upon them; much of it throughout the history of the world can be traced back to religious interpretations.

Women in the Civil War

Bibliography

Boehnke, Jan, *Heroines of the American Civil War*, Vol. 8, No. 1, The Ozarks Reader Magazine, Neosho, Missouri, 2011.

Funk & Wagnalls New Encyclopedia, Funk & Wagnalls, Inc., New York, 1979.

Goldfield, David, *The American Journey: A History of the United States,* Prentice Hall, Upper Saddle River, New Jersey, 1998.

Guitteau, William B., *The History of the United States*, Houghton Mifflin Company, 1942.

Halleck, Reuben Post, *History of Our Country*, American Book Company, 1923.

Holcombe & Adams, *An Account of the Battle of Wilson's Creek or Oak Hills*, Dow & Adams Publishers, Springfield, Missouri, 1883.

Roak, James L., *The American Promise: A History of the United States*, Bedford Books, Boston, 1998.

Rowland, Tim, *Strange and Obscure Stories of the Civil War*, Skyhorse Publishing, Inc., New York, 2011.

Sullivan, Walter, *The War the Women Lived: Female Voices from the Confederate South*, J.S. Sanders & Company, Nashville, 1995.

Civil War Food and Rations

AT A TIME before modern food preserving techniques were discovered, a Civil War soldier's food options were simpler and somewhat limited. Much of a soldier's diet often consisted of foraging the countryside for whatever was available—from nature or planted. One Confederate soldier wrote that: "Just in front of our position was a small field of corn, and, as we needed some for our horses and ourselves—for parched corn was our chief provender at this time—when there was a lull in the firing several of us crawled on our hands and knees, trailing our forage-sacks, and reached the corn rows without being seen. The watchful enemy, observing the tops of the stalks in commotion, turned loose on us so effectively that the corn detail suspended operations until night-time." Foragers that were sallied forth looking for sustenance found wild fruit, nuts, berries, greens, game, and many other things.

Union troops bivouacking, or while on the battlefield, often consumed salt pork, coffee, sugar, salt, vinegar, dried fruit or vegetables, beans, and hardtack; while Southerners had rations like bacon, cornmeal, tea, sugar or molasses, and fresh vegetables. One main staple was hardtack, because it was cheap to make and could be carried for long periods of time. However, hardtack could become so hard overtime that the troops had to use stones, or the butt of their rifles, to pulverize the bread for consumption. A number of soldiers broke and cracked their teeth trying to eat hardtack! It came to be known as "teeth dullers," "sheet-iron

CURIOUS AND UNUSUAL CIVIL WAR STORIES

Reenactment camp photo taken in Bentonville, Arkansas.

Civil War Food and Rations

crackers," "ship's biscuit," and "flour tile," among other things. Sometimes they would soak the hardtack in coffee, water, stew, or fry it in pork fat, in order to soften it enough to eat it. It was also often compared to eating wall plaster. Another common tale of woe concerning hardtack was when it became infested with bugs or maggots—hence, the name "worm castles." Soldiers often broke it open to find live worms.

Civil War soldiers also resorted to eating rats, or even dogs. One soldier who spent time in Camp Morton which was located near Indianapolis, Indiana, recalled that: "All rats which could be trapped were eaten, and to my knowledge one fat dog was captured by my messmates, cooked, and eaten. I was invited to partake; and, although the scent of the cooking meat was tempting, I could not so far overcome my repugnance to this animal as an article of diet as to taste it."

There were other reports of eating rats and how difficult it was to catch them, as they sometimes examined traps and showed that they are smarter at times than can be imagined by humans. At hospitals they were known to eat the poultices which were applied to patients, and made off with bran-stuffed pads that had been placed under the limbs of injured soldiers.

A rat recipe from the period gives the creature a chance for some future customers, and claims that it should at first be skinned, gutted and cleaned, beheaded, and laid upon a board. The legs should then be stretched out tight and tacked down; then, basted with bacon fat and roasted until well done.

Another favorite dish during the War of the Rebellion was called "coosh," which was made from bacon fried with cornmeal and water. The "Johnnie cake" made of cornmeal was also a popular choice for hunger soldiers—possibly, even better than roasted rat!

Bibliography

Botkin, B.A., *A Civil War Treasury of Tales, Legends and Folklore*, Random House, New York, 1960.

Jackson, Rex T., *Civil War Rations*, Vol. 50, No. 1, The Ozarks Mountaineer Magazine, Kirbyville, Missouri, 2002.

Wyeth, John Allan, *With Sabre and Scalpel: The Autobiography of a Soldier and Surgeon*, Harper & Brothers Publishing, New York and London, 1914.

Titanic Survivor Wrote About the Civil War

COLONEL Archibald Gracie, who wrote *The Truth About Chickamauga*, spent about seven years studying the historic battle and tramping over the battleground taking measurements and analyzing the numerous movements and positions of the troops. But in April 1912, Gracie was aboard the doomed *RMS Titanic*, a grand ocean-liner launched on May 31, 1911, from the Harland & Wolff Shipyard in Belfast, Ireland—the "Emerald Isle." For about ten months the great 882-feet long ship was outfitted with unparalleled appointments, in preparation for its maiden voyage. The *Titanic* departed its final stop at Queenstown (now Cobh), Ireland, on April 11, 1912, on its way to America and New York City. On the night of April 14, 1912, the ship sideswiped an iceberg and foundered at 2:20 a.m. on April 15—over 1,500 souls went to a watery grave.

On the 14th, however, passenger Isidor Straus, co-owner of Macy's—a budding store in New York, had just finished reading Gracie's Civil War book about Chickamauga, which Gracie had loaned him. The weary author had taken the overseas trip to recuperate from the work of writing the book; and somewhere along the way he had given Straus the book to pass the time while abroad the gigantic steamer.

Isidor Straus and his beloved wife, Ida, who chose to remain on board the ill-fated ship with her husband, became casualties. Gracie

Titanic under construction.
(Photo taken at the Titanic Museum in Branson, Missouri.)

Titanic Survivor Wrote About the Civil War

would survive and in the next few months he would author another book, *The Truth About the Titanic*; however, he would not live to see it published—Archibald Gracie died on December 4, 1912, and his book became available the very next year. Apparently, his health had deteriorated after the *Titanic* ordeal as a result of his exposure in the frigid Atlantic waters.

Bibliography

Gracie, Archibald, *The Truth About the Titanic*, Sutton Publishing, 1913.

Tibballs, Geoff, *The Titanic: The Extraordinary Story of the "Unsinkable" Ship*, Reader's Digest, 1997.

Wyeth, John Allan, *With Sabre and Scalpel: The Autobiography of a Soldier and Surgeon*, Harper & Brothers Publishers, New York and London, 1914.

Civil War Love Feast: Compassion and Mercy

DURING THE Civil War, prisoners of the 3rd Ohio Regiment on their way to a Confederate prison in Richmond, Virginia, were traveling through Tennessee and made a stop. As a result, some curious soldiers of the 54th Virginia heard about the Union prisoners and wanted to go see them in their hopeless, captive condition to gloat at their misfortune. When the rebels finally arrived and saw them in their sad plight, things didn't happen as they had first imagined, their hearts went out to them and so they scurried back to their camp to tell others and to get food and sustenance for the weary prisoners. Before long, they returned to them with a meal fit-for-a-king—so deeply touched by their Southern hospitality and mercy they vowed never to forget it as long as they lived.

By and by the Union prisoners were exchanged and they returned to duty, some of them were camped near Kelly's Ferry on the Tennessee River. It just so happened that after the Battle of Missionary Ridge on November 25, 1863, a large number of Confederate soldiers of the 54th Virginia were captured and taken to Kelly's Ferry where they would eventually be transported to prison. Ironically, a soldier of the 3rd Ohio inquired about the prisoners and what company they were with, and upon learning that it was some of the 54th Virginia, he immediately informed his comrades who quickly gathered together a banquet of food and other things to

Idle muskets at a Civil War reenactment in Lexington, Missouri.

provide to the prisoners. Their reunion was a sight to behold as they again shared a memorable, merciful feast—putting aside their weapons and joining in on a special moment in history that neither side would ever forget, or regret.

Bibliography

Botkin, B.A., *A Civil War Treasury of Tales, Legends and Folklore*, Random House, New York, 1960.

Grant, Ulysses S., *Chattanooga*, Battles and Leaders of the Civil War, The Century Company, 1887.

Civil War Aeronautics: Silk Balloons

EXPERIMENTING with balloons during the American Civil War to reconnoiter enemy territory and positions was utilized by both the North and the South. Floating over and behind enemy lines using a large silk bag filled with gas which was lighter than air, was used successfully on a number of occasions.

Ballooning was tried as early as 1783 by two well-to-do papermakers of Annonay, France, Jacques Etienne Montgolfier and Joseph Michel Montgolfier, who experimented with hot air balloons. The first balloon to fly in America, though, was attempted in Philadelphia, Pennsylvania, about 10 years later on January 9, 1793.

On July 1, 1859, the first longest, recorded balloon flight of about 809 miles in just under 20 hours time, was accomplished by aeronauts John Wise and John LaMountain. The 130-feet high and 60-feet in diameter *Atlantic* was launched from Washington Square in St. Louis, Missouri, along with balloon enthusiast O.A. Gager and reporter William Hyde; Gager had funded the construction of the balloon. In case the *Atlantic* came down in water, due to its projected path, it had a wooden boat suspended below the basket.

The brave, aeronautical pioneers reached Lake Erie near Sandusky, Ohio, the day after its launch; they made it to the Buffalo, New York and Niagara Falls area around mid-day. As they were floating above and over Lake Ontario a strong storm and downdraft forced the *Atlantic* to descend in spite of their best efforts to stay airborne. Somehow, reaching land the boat dangling below began to drag and smash into the trees until the balloon was torn asunder, which ended Wise's and LaMountain's historic flight near Henderson, New York.

When the War Between the States broke out, Wise and LaMountain sought positions in the Union Balloon Corps; however, the honor eventually fell to Professor Thaddeus S.C. Lowe. The difference that existed between Wise and Professor Lowe was in free or tethered fight.

Making a new record journey on April 20, 1861, from Cincinnati, Ohio, to South Carolina—about 900 miles distance, Professor Lowe was able to showcase to the military and the world the value of ballooning. During the Battle of Seven Pines, for instance, Professor Lowe observed the action from his gas balloon the *Intrepid* on the north side of Chickahominy River. On occasion, the Confederates would attempt to shoot balloons down but without success. In the *Official Records of the Union and Confederate Armies*, Professor Lowe is praised for his groundbreaking work and contributions: "To Professor Lowe, the intelligent and enterprising aeronaut, who had the management of the balloons, I was greatly indebted for the valuable information obtained during his ascensions."

Confederate General James Longstreet also felt the need to comment about the use of balloons during the Civil War and how the South could have benefited from them, had they enough war chest to afford it. He saw how the Union was able to float gracefully into the air beyond the reach of shot and shell and gather important intelligence. Reminiscing, Gen. Longstreet recalled a time when someone came up with the "bright" idea of gathering together silk dresses from Southern "belles" to sew them together and create a balloon—which they did. During the Seven Days' campaign waged from June 25 to July 1, 1862, the South launched their own version of a balloon that had the honor and distinction of looking like a flying quilt.

The only place available for them to get gas for the balloon was

Civil War Aeronautics: Silk Balloons

Author's sketch of Professor Thaddeus S.C. Lowe's *Intrepid*.

in Richmond, Virginia—the Confederate capital, so it had to be filled there and moved by securing it to a locomotive or a riverboat steamer in order to get it by rail or river to where it was going to be used. The patchwork balloon was eventually confiscated by the Federals on the James River after the steamboat that was transporting it became lodged on a sandbar and was unable to free itself. Gen. Longstreet was unhappy to see the precious silk-made balloon of donated Southern dresses lost to Northern hands.

Civil War Aeronautics: Silk Balloons

Bibliography

Botkin, B.A., *A Civil War Treasury of Tales, Legends and Folklore*, Random House, New York, 1960.

Funk & Wagnalls New Encyclopedia, Funk & Wagnalls, Inc., New York, 1979.

Legrand, Jacques, *Chronicle of America*, Chronicle Publications, Mount Kisco, New York, 1989.

Official Records of the Union and Confederate Armies, Washington: Government Printing Office, 1881.

Porter, John Fitz, *Hanover Court House and Gaine's Mill*, Battles and Leaders of the Civil War, The Century Company, 1887.

Young, Joshua, *Myths and Mysteries of Missouri: True Stories of the Unsolved and Unexplained*, Globe Pequot Press, 2014.

Losing the *Indianola*: The Audacity and Trickery of Admiral Porter

OF ALL the curious and unusual military tactics that occurred during the American Civil War, one clever tale, however bizarre and strange as it might have been, actually worked well and achieved its goal. The crazy occurrence happened in the Vicksburg, Mississippi vicinity on the Mississippi River.

A Union steamer, the *Indianola*, had left the Red River and was retracing its steps north on the Mississippi. Meanwhile, the *Queen of the West*, a Union ram which had recently been raised, repaired, and put into service by the Confederates, along with the ram *Webb*, which had been in the Red River for some time, and a couple of smaller cotton-clad vessels were giving pursuit to the *Indianola*. On February 24, 1863, the Union gunboat was overtaken by the hostile Confederate flotilla just north of the Grand Gulf and a short distance south of Warrenton, Mississippi (some sources give the location on the shores of the Jefferson Davis Plantation). In any case, a sharp river-battle which lasted for about an hour and a half was waged. In the end, the *Indianola* was rammed seven times leaving a gaping hole and disabling its steering mechanism which eventually forced it ashore where it sank and settled to the bottom in about 10 feet of water. The men of the *Indianola* did their utmost to throw overboard everything they could to keep it from being confiscated by the enemy; while the Confederates did the opposite—they boarded the

CURIOUS AND UNUSUAL CIVIL WAR STORIES

Indianola to take prisoners and anything of value that they could utilize in the war effort.

Now this is where it gets good: Not long after the sinking of the *Indianola*, Union Admiral David D. Porter, upon hearing about yet another vessel falling into enemy hands and at risk of being raised and added to the Confederate arsenal, came up with a rather uncommon military solution to the dilemma—he would build a giant "dummy" ironclad. The plan was to float it downstream past Vicksburg and toward the *Indianola* in order to fool and beat the greedy treasure-hunters into a hasty retreat—and it had to be done quickly!

Admiral Porter gathered together a number of carpenters, workers, and needed supplies to convert a coal barge into a formidable-looking vessel of war. The dummy, 300-foot monitor was completed with a wooden deck and sides, two wheelhouses, log casemate with "Quaker" guns (fake log cannons) sticking out, two old boats on davits, a couple of smokestacks made of recycled hogsheads (pork barrels), and a flag flopping in the breeze from its patriotic jack-staff. To make the whole ruse look authentic, a pot of burning tar and oakum (hemp or jute twisted together) was placed inside each smokestack to give the appearance that it was indeed under a head of steam and in operation. When all was made ready, the hoax-boat was hauled out into the Mississippi currents and released just upriver from Vicksburg's shoreline batteries.

As Admiral Porter's fiendish creature was smoking, belching, and drifting past Vicksburg, a deafening din of cannon-fire broke the stillness of the night. The Confederate batteries gave it all they could muster, but somehow the great river beast was unstoppable and cruised past and onward towards what was left of the *Indianola*. Meanwhile, couriers were dispatched to warn the Southern flotilla that a Union gunboat was on the way. As a result, the *Queen of the West*, *Webb*, and cotton-protected boats directed their attention away from salvaging the *Indianola* to the oncoming threat. Not wanting a confrontation from such a possible foe as was reported, they torched the *Indianola* and fled downstream in all haste. As it so happened, Admiral Porter's brainchild had run aground a couple of miles above them—apparently, the thing had no stomach for a fight.

Losing the Indianola

When the Confederates in the river and upon the shores learned of the audacity and trickery of the Union admiral, they were furious, embarrassed, and awed by the whole affair. Even though it never became a widely-known Civil War incident, it did stop the Confederacy from raising and refurbishing another Union boat—a real one.

Bibliography

Botkin, B.A., *A Civil War Treasury of Tales, Legends and Folklore*, Random House, New York, 1960.

Coombe, Jack D., *Thunder Along the Mississippi: The River Battles that Split the Confederacy*, Bantam Books, 1998.

Greene, Francis Vinton, *Campaigns of the Civil War: The Mississippi*, 1882.

Soley, James Russell, *Naval Operations in the Vicksburg Campaign*, Battles and Leaders of the Civil War, The Century Company, 1887.

John Charles Fremont: "Pathfinder of the West"

ONE NINETEENTH century American icon that earned the title of the "Pathfinder of the West," who spent most of his lifetime as an explorer, soldier and politician in the frontier, was John Charles Fremont. Fremont was born in Savannah, Georgia in 1813 and attended college in Charleston, South Carolina. He taught mathematics and worked as an assistant engineer, and by 1838 he had been commissioned a second lieutenant as an Army Engineer. The next year he joined Joseph Nicolas Nicollet, a Frenchman, and the Army Topographical Corps on a scientific expedition into the plateau area between the Upper Mississippi and Missouri Rivers to survey and map the region. During the expedition, Fremont benefited from Nicollet's experience and scientific knowledge.

In 1840 Fremont became so enthralled with the daughter of Missouri Senator Thomas Hart Benton, Jessie Benton, that despite Sen. Benton's robust attempt to separate them, they secretly married in 1841. Sen. Benton would eventually "warm up" to the union and went on to support his new son-in-law by backing his frontier ventures.

Not long after the "honeymoon," in 1842, Fremont headed off to map the Oregon Trail. Along the way he gathered important information about the region's terrain and climate and ascended a 13,730-foot high mountain peak in what is now present-day

Illustration of the California Bear Flag which was raised in northern California on June 14, 1846, and lowered on July 11, 1846; it was replaced by the flag of the United States.

John C. Fremont

Wyoming—the second-highest peak in the Wind River Mountain Range which came to be known as "Fremont Peak." The reports he made about his expedition helped to encourage and inspire settlement and western expansion. His first success was followed by a second journey the very next year in 1843, when Fremont finished mapping the Oregon Trail—trekking over the Sierra Nevada Mountains and to the mouth of the Columbia River and the Pacific coast in present-day California. Along for the adventure was pathfinder and scout Christopher "Kit" Carson, who guided them to the frontier settlements in Oregon and northwest Nevada where they spent the winter at Sutter's Fort before reaching the California coast. On their return trip they crossed the southern Sierra Nevada Mountains and the Great Salt Lake. On August 7, 1844, Fremont published his account and findings of his experience in the West while back in St. Louis, Missouri.

Sutter's Fort (Sacramento, California), located near the Sacramento River, gained historical notoriety when gold was found at John Sutter's sawmill on the bank of the American River at Coloma. The lucky discovery was made by James Marshall, the mill's builder and foreman. On January 24, 1848, the great gold rush began when Marshall spied the sparkling, precious metal dust in the millrace. The "strike" was kept secret for a short time but it didn't last long, a Mormon by the name of Sam Brannan who planned to build a business near the mill to cater to the needs and whims of the miners, held up a bottle filled with gold and barked out the breaking news to the San Francisco public—the rush was on! The strike also spawned a culture of lawlessness—crimes of murder, highway robbery and cattle rustling to name a few. One outlaw and desperado during the gold rush era was Joaquin Murrieta, who was dealt with and killed on July 25, 1853. In order to discourage further such activities, however, Murrieta's head was put on public display in a large jar throughout the state.

Fremont returned to California for a third expedition in 1845 to explore the Great Basin and the Western coastline. While in California the Mexican War (1846 to 1848) broke out and Fremont was commissioned major and commanded troops against Mexico. On July 7, 1846, Fremont, along with sixty buckskin-clad frontier soldiers, led the "Republic of California" during the "Bear Flag Revolt," raising the Bear Flag at Monterey. The flag was made from a flour sack which had a star, grizzly bear and the words "California Republic" painted on it. When some of the Californians were declaring their independence from Mexico, they cried out that they were just playing the "Texas game." After their "victory-dance," the local residents declared Fremont the head of their new Republic. The Bear Flag flew proudly in the breeze but was lowered a few days later on July 11, 1846, and replaced by the "Stars and Stripes"—the flag of the United States of America. About the conquest of New Mexico and California during this time, one of Fremont's buckskin-clad soldiers had this to say: "We...marched all over California from Sonoma to San Diego, and raised the American flag without opposition...."

Following the American conquest and victory over Mexico, Fremont was appointed California civil governor by American naval commander Commodore Robert Field Stockton; however, a dispute arose over authority between Commodore Stockton and American Brigadier General Stephen Watts Kearny.

In Gen. Kearny's earlier days he visited settlements in the frontier and into the untamed West. Captain Kearny commanded Fort Crawford at Prairie du Chien (Wisconsin) on the Upper Mississippi River; while there, he entertained men like Zachary Taylor and Jefferson Davis. Lieutenant Colonel Kearny was the first commander at Jefferson Barracks in St. Louis, Missouri; the barracks was established in 1826 as an Infantry School of Practice and hosted such historical figures as Ulysses S. Grant, William Tecumseh Sherman, Winfield Scott Hancock, Zachary Taylor, Robert E. Lee, James Longstreet and Jefferson Davis. During the Civil War over 200 generals served at the barracks. While there, Lieut. Col. Kearny commanded the first cavalry unit and gained the distinction of being the "Father of the U.S. Cavalry," but he eventually migrated to California where he built a number of army posts and served in the Mexican War. By 1848, General Kearny had achieved the rank of major general.

As for Gov. Fremont, his disobedience to Gen. Kearny's orders earned him a charge of mutiny, prejudicial conduct and

John C. Fremont

insubordination, which was rewarded by arrest and court-martial. As a result, President James Knox Polk dropped the charges against him and Fremont resigned from the army. Fremont quickly rebounded and went to work searching for possible mountain passes for a rail line to run from the Rio Grande River to California. His popularity in California never faded, however, and in 1850 he was elected as one of two to serve as senator. Before long, in 1856, the dashing adventurer, John C. Fremont, who was, ironically, an antislavery Southerner, was nominated as the Republican candidate for President of the United States. During this time his wife Jessie was his greatest asset, who knew Washington politics as well as he knew the Wild West. His opponents were James Buchanan and Millard Fillmore. The election was a "nail-biter"—Buchanan won 174 electoral votes (1,833,000 popular votes), Fremont won 114 electoral votes (1,339,000 popular votes), and Fillmore lagged behind with 8 electoral votes (872,000 popular votes).

When the American Civil War erupted, Fremont was not forgotten; he was placed in command of the Department of the West and stationed in St. Louis, Mo. In the *Official Records of the Union and Confederate Armies*, on July 3, 1861, the War Department in Washington made it official when the Adjutant General's Office wrote: "The State of Illinois and the States and Territories west of the Mississippi River and on this side of the Rocky Mountains, including New Mexico, will in future constitute a separate military command, to be known as the Western Department, under the command of Major General Fremont, of the U.S. Army, headquarters at Saint Louis."

Fremont's antislavery tendencies eventually got him into deep trouble, after he declared martial law and ordered the emancipation of Missouri slaves and seized Confederate property—abolitionists were delighted with Fremont. About the daring order, Fremont wrote in the *Official Records* on August 14, 1861: "I hereby declare and establish martial law in the city and county of Saint Louis. Maj. J. McKinstry, U.S. Army, is appointed provost-marshal. All orders and regulations issued by him will be respected and obeyed accordingly."

At the time, President Abraham Lincoln was not amused; he

disagreed with Fremont's orders and removed him from office. Lincoln preached in his inaugural address that he had no reason to "interfere" with slave states. When Fremont relinquished his command of the Western Department to Major General D. Hunter on November 2, 1861, he addressed the Soldiers of the Mississippi Army and said: "Agreeably to orders this day received I take leave of you. Although our army has been of sudden growth, we have grown together, and I have become familiar with the brave and generous spirit which you bring to the defense of your country, and which makes me anticipate for you a brilliant career. Continue as you have begun, and give to my successor the same cordial and enthusiastic support with which you have encouraged me. Emulate the splendid example which you have already before you, and let me remain, as I am, proud of the noble army which I had thus far labored to bring together.

"Soldiers, I regret to leave you. Most sincerely I thank you for the regard and confidence you have invariably shown to me. I deeply regret that I shall not have the honor to lead you to the victory which you are just about to win, but I shall claim to share with you in the joy of every triumph, and trust always to be fraternally remembered by my companions in arms."

Once again in 1864 John C. Fremont, the military hero, became the Radical Republican faction nominee for President of the United States; however, after President Lincoln agreed to remove Montgomery Blair from his cabinet as Postmaster General, Fremont withdrew his bid for the "White House."

After the War of the Rebellion ended in 1865, Fremont squandered his savings on a transcontinental railroad venture, but the tide changed again for him in 1878 when he was named as the territorial governor of Arizona—he served until 1883. Before he died in 1890, he had regained his army status as major general.

From explorer to soldier, to a politician with one foot in the White House, John C. Fremont's contribution to American history solidifies his place with other important icons of the West. Although his name is forever associated and connected with "the Pathfinder,"—he was, undoubtedly, so much more.

John C. Fremont

Chronology

1813—John Charles Fremont was born on Savannah, Georgia.
1838—Fremont was commissioned a second lieutenant as an Army Engineer.
1841—Fremont marries Senator Thomas Hart Benton's daughter, Jessie Benton.
1842—Fremont heads out to map the Oregon Trail.
1843—Fremont leaves on his second expedition to finish mapping the Oregon Trail, also known as the "California Trail" and the "Great Western Road,"—and by Native Americans as the "Great Medicine Road."
1844—Fremont publishes his findings while back in St. Louis, Mo.
1845—Leaves on his third expedition to explore the Great Basin and Pacific coast.
July 7, 1846—Fremont leads the Republic of California during the Bear Flag Revolt and raises the Bear Flag.
July 11, 1846—The Bear Flag is lowered and the flag of the United States is raised in California.
1850—Fremont becomes a senator of California.
1856—Fremont is nominated as the Republican candidate for President of the United States but is defeated by James Buchanan.
July 3, 1861—Fremont is placed in command of the Department of the West during the American Civil War.
August 14, 1861—Fremont declares martial law in St. Louis, Missouri.
November 2, 1861—Fremont relinquishes his command of the Western Department.
1864—John C. Fremont named presidential nominee for the Radical Republican faction but eventually withdraws from the race.
1878—Becomes territorial governor of Arizona and serves until 1883.
1890—John Charles Fremont dies.

CURIOUS AND UNUSUAL CIVIL WAR STORIES

Bibliography

Amsler, Kevin, *Final Resting Place : The Lives and Deaths of Famous St. Louisans*, Virginia Publishing Company, St. Louis, Missouri, 1997.

Athearn, Robert G., *American Heritage New Illustrated History of the United States*, Vol. 6, *The Frontier*, Fawcett Publications, Inc., One Astor Plaza, New York, N.Y., 1971.

Chronicle of America, Chronicle Publications, Inc., Mount Kisco, New York, 1989.

Faragher, John Mac; Buhle, Mari Jo; Czitrom, Daniel; Armitage, Susan H., *Out of Many*, Prentrice Hall, Upper Saddle River, New Jersey, 1997.

Funk & Wagnalls New Encyclopedia, Funk & Wagnalls, Inc., New York, 1979.

Golay, Michael; Bowman, John S., *North American Exploration*, John Wiley & Sons, Inc., Hoboken, New Jersey, 2003.

Guitteau, William Backus, *The History of the United States*, Houghton Mifflin Company, The Riverside Press, 1942.

Halleck, Reuben Post, *History of Our Country*, American Book Company, New York, 1936.

Havighurst, Walter, *Voices on the River*, Simon & Schuster, New York, N.Y., 1964.

Official Records of the Union and Confederate Armies, Washington: Government Printing Office, 1881.

Roark, James L.; Johnson, Michael P.; Cohen, Patrick Cline; Stage, Sarah; Lawson, Alan; and Hartmann, Susan M., *The American Promise: A History of the United States*, Bedford Books, Boston, 1998.

Old Creek Chief Hopoeithleyohola: Civil War Refugees

GREED AND avarice blossomed under the "Stars and Stripes" in the early 1800s, as America snatched the eastern homelands from Native Americans. For many years the original inhabitants faced genocide and decimation—a sinful stain and mark upon Christian society and American history. Each and every hollow treaty and promise left many Native Americans with less and less, while the nation waxed rich at their expense. Concerning the Creeks, by 1827 they had ceded all their remaining property in Georgia to the whites and reluctantly agreed to relocate to Indian Territory (Oklahoma). Like sheep they were herded to the West by the thousands through ruff, rugged, and untamed wilderness. When winter hit, they endured snow and death-dealing frigid, arctic air while they were forced to press on scantly clothed and, in many cases, barefooted. Finally, they reached Fort Gibson (Oklahoma), their new homeland chosen for them, and saw "Old Glory" flapping in the breeze above it. The Nations soon filled up with many and various tribes, and they began to rebuild their lives there.

When the War Between the States began in 1861, the North and South vigorously sought to bog Native Americans down in the nation's struggle—over preserving the Union and the sin of human slavery. A number of tribes, especially the Cherokee, Creek, Chickasaw, Choctaw, and Seminole participated to some degree.

CURIOUS AND UNUSUAL CIVIL WAR STORIES

In 1839 Reverend Jesse Bushyhead settled in northeast Oklahoma (Indian Territory) after the Cherokee Removal. A Baptist Mission was established by 1841. The site also served as a ration station and came to be known as Breadtown, or *Ga-Du-Hv-Ga-Du*. The Cherokee *Messenger* was printed at this site beginning in 1844; it was the first periodical published in Oklahoma. The Mission was torched during the Civil War because of the anti-slavery message taught by the missionaries. The present church building was built in 1888.

Old Creek Chief Hopoeithleyohola

One of the first to get involved was the old Creek Chief Hopoeithleyohola (*Hok-tar-hah-sas-Harjo*), also known as Opothleyoholo or Opothle Yahola. Hopoeithleyohola leaned in favor of the Union, but in an attempt to sway his loyalties, Confederate Colonel Douglas H. Cooper, a former Indian agent, made several unsuccessful attempts to meet with him. Instead of conversion, Hopoeithleyohola gathered together a pro-Union force of loyal Creeks and various other allies north of the Cimarron River in the Creek Nation. As a result, a number of hotly contested battles in Indian Territory followed: Round Mountain (Red Forks), November 19, 1861; Bird Creek (Chusto-Talasah), December 9, 1861; and Patriot Hills (Chustenahlah), December 26, 1861. The fighting did not go well for Hopoeithleyohola and his loyalists; they were forced to retreat towards southern Kansas. By one account they fled in "...scattered lines, with hosts of stragglers, the enfeebled, the aged, the weary, and the sick, they crossed the Cherokee Strip and the Osage Reservation and, heading steadily towards the northeast, had finally encamped on the outermost edge of the New York Indian Lands, on Fall River, some sixty odd miles west of Humboldt, [Kansas]."

As early as September 10, 1861, E.H. Carruth, Commissioner of the U.S. Government, wrote to Hopoeithleyohola and made these promises: "I am authorized to inform you that the President will not forget you. Our Army will soon go South, and those of your people who are true and loyal to the Government will be treated as friends. Your rights to property will be respected...the President is still alive. His soldiers will soon drive these men who have violated your homes from the land they have treacherously entered...."

Indians from the Nations wandered for weeks in search of some sort of sanctuary from the American war—the result was not good. "It was inconceivably horrible. The winter weather of late December and early January had been most inclement and the Indians had trudged through it, over snow-covered, rocky, trailess places and desolate prairie, nigh three hundred miles. When they started out, they were not any too well provided with clothing; for they had departed in a hurry, and, before they got to Fall River, not a few of them were absolutely naked."

What followed was a sad and appalling tale of woe and death; as many as 7,600 refugees had no other recourse than to wait for the mercy of spring to come and rescue them. However, by the time their savior would arrive, there would be untold pain and suffering. In great need of help, the loyal Hopoeithleyohola drafted a letter to President Abraham Lincoln and sent it to Fort Leavenworth, Kansas, dated January 28, 1862, which read: *"To our Great Father, the President of the United States*: FATHER: We are told by our friends that there is some doubt as to whether the great war chief, General [James Henry] Lane, will command the expedition to our country.

"Our object in having this letter sent to you is to beg that General Lane be placed in command of that expedition, as we believe no warrior can place us in possession of our country again as effectually as he can. Our people have heard of General Lane many seasons ago. They have heard how with but a handful of warriors he beat back the enemy when they were as numerous as the leaves of the forest and restored peace and quiet to Kansas.

"Our people have been told that he would come with an army to restore them to their homes and to avenge the great wrongs they have suffered.

"It has made their hearts glad to hear it.

"Our people have suffered a great deal. They have been driven from their homes in the dead of winter when the earth was clothed with white. Many of them have frozen to death. All of them have lost all they possessed.

"There are now 6,000 women and children in Southern Kansas without tents, but scantily clothed, and exposed to all the horrors of a severe winter.

"Our agents have done and are doing all they can to relieve us, but we leave comfortable homes in our own country and we wish to be restored to them.

"General Lane is our friend. His heart is big for the Indian. He will do more for us than any one else. The hearts of our people will be sad if he does not come. They will follow him wherever he directs. They will sweep the rebels before them like a terrible fire on the dry prairie.

Old Creek Chief Hopoeithleyohola

"We beg our Great Father and our great war chief, General McClellan, that they will listen to the prayers of their children."

As time dragged on and relief never arrived, a surgeon, Dr. A.B. Campbell, was sent to make an assessment of the refugee camp. Dr. Campbell saw that, in many cases, their only protection and shelter from the driving wind and snow was the abundant prairie grass and the bits and pieces of rags and other things they had to stretch upon twigs and branches. He witnessed their need for medical attention as they suffered from inflammatory conditions—chest, throat, and eye. Many of the refugees had lost their toes due to the extreme cold weather. "Dead horses were lying around in every direction and the sanitary conditions were so bad that the food was contaminated and the newly-arriving refugees became sick as soon as they ate."

Much of the food that had been doled out for them had been rejected by the military as unfit to eat. The demand for sustenance and medicine far outweighed the available supply on hand.

Agents George C. Snow and William G. Coffin reported on February 13, 1862, that "...destitution, misery and suffering amongst them is beyond the power of any pen to portray, it must be seen to be realized—there are now here over two thousand men, women, and children entirely barefooted and more than that number that have not rags enough to hide their nakedness, many have died and they are constantly dying."

Springtime was the only helper to finally rear its head and offer a warm embrace—at least to those still living that could enjoy or benefit from it. The plight of Native Americans during the Civil War was grim, as Indian Territory became a wasteland of burned-out homes and schools—not much to go back to. Thousands of them fought in the war and, in some cases, against each other under Confederate and Union flags. Hopoeithleyohola never returned to the battlefield, but for the remainder of his life he, undoubtedly, never forgot the hollow and empty promises of aid and friendship offered by the Federal government which never materialized. It left small consolation to the many graves of loved ones littering the Kansas prairie.

Bibliography

Abel, Annie Heloise, *The American Indian as Participant in the Civil War*, Arthur H. Clark Company, Cleveland, 1919.

Jackson, Rex T., *A Trail of Tears: The American Indian in the Civil War*, Heritage Books, Inc., Westminster, Maryland, 2004.

Jahoda, Gloria, *The Trail of Tears*, Holt, Rhinehart and Winston, New York, 1975.

Mooney, James, *Myths of the Cherokee*, Nineteenth Annual Report of the Bureau of American Ethnology to the Secretary of the Smithsonian Institution, 1900.

Official Records of the Union and Confederate Armies, Washington: Government Printing Office, 1881.

Nathan Bedford Forrest: "When Churchyards Yawn"

AFRICAN AMERICANS during the Civil War were later, after having proved their battle-character, considered fierce fighters—and for good reason. In many cases they received no quarter from the Southerners, but were killed instead of being taken prisoner.

At Fort Pillow, Tennessee, on April 12-13, 1864, Confederate troops under command of Major General Nathan Bedford Forrest, reportedly, massacred 238 African American soldiers of Major Lionel F. Booth's 6th United States Heavy Artillery. About three-quarters of Major Booth's 557-man army were slaughtered, including some of the 13th Tennessee Cavalry and even a number of women and children. It was reported that while General Forrest was demanding that the fort surrender under his flag of truce, a number of his troopers slipped into the fort, and when the Federals refused to surrender, the Confederates attacked while shouting "No quarter." Eyewitnesses reported that many of them were begging for mercy while they were being shot and bayoneted—the bloody work continued through the night and again the next morning; Major Booth was also one of the casualties of the Fort Pillow massacre. The news of the battle spread like wildfire in the North and prompted an investigation by the Committee on the Conduct of the War, which charged the Confederates with "an indiscriminate slaughter, sparing neither age nor sex, white or black, soldier or civilian."

In defense of the action at Fort Pillow, General S. D. Lee,

Author's sketch of a Ku Klux Klan Robe and Hood.

Nathan Bedford Forrest

Confederate commander of the Department of Alabama, Mississippi, and East Louisiana, had this to say: "...fearful results are expected to follow a refusal to surrender.

"The case under consideration is almost an extreme one. You had a servile race armed against their masters, and in a country which had been desolated by almost unprecedented outrages. I assert that our officers, with all the circumstances against them, endeavored to prevent the effusion of blood, and as an evidence of this I refer you to the fact that both white and colored prisoners were taken...."

Gen. Forrest also reported on the attack of Fort Pillow, a Union station for the purpose of protecting the navigation of the Mississippi River, and said: "As our troops mounted and poured into the fortification, the enemy retreated towards the river, arms in hand, and firing back, and their colors flying; no doubt expecting the gun-boat to shell us away from the bluff and protect them until they could be taken off or [reinforced]. As it was, many rushed into the river and were drowned, and the actual loss of life will perhaps never be known, as there were quite a number of refugee citizens in the fort, many of whom were drowned and several killed in the retreat from the fort."

Gen. Forrest reported that they captured 164 Federals, 75 African American troops, and about 40 African American women and children, and "after removing everything of value...the warehouses, tents, etc., were destroyed by fire."

General Ulysses S. Grant wrote that "Forrest made a report in which he left out the part which shocks humanity to read."

As to further character-damage and proof, on December 24, 1865, only a few months after the end of the Civil War, Nathan Bedford Forrest and six other ex-Confederate officers organized the secret, terrorist society known as the Ku Klux Klan (KKK); its name was adapted from the Greek word *kuklos* meaning "circle." Established in Pulaski, Tennessee, the "Klan" was active in the South during Reconstruction. The Klansmen "wore a white mask, high cardboard hats, and a long gown. They rode forth at midnight, 'when churchyards yawn,' on masked horses shod with felt." Their object was to stop or hinder African Americans from voting or getting involved with politics, and to keep them more subservient to whites. The Klan would leave threatening and "Mysterious warnings decorated with skull, coffin, and crossbones...to individuals. If the warning was not heeded the individual was sometimes whipped, maimed, or killed." Hiding behind these masks were Americans who were racists, prominent community leaders, Christian ministers and Protestant Holy Rollers, bigots, and just about every vigilante-type scumbag that felt the need to enforce their sinful nature upon their fellowman—whipping, lynching, and cutting throats.

Eventually, however, the white supremacists got so out-of-control that even Forrest had to distant his self from the hit-and-run guerrilla bands who were on the loose scouring the hillsides for their next poor victims. One of their favorite things to do was to burn fiery crosses near the homes of their targets. They also worked to force Catholics, Jews, and foreigners from society while they believed that they were defending traditional, conservative values—mostly based on their own misguided Bible research. As a result, Grand Wizard Forrest disbanded the KKK in 1869—many citizens were undoubtedly relieved. But not to be denied, a new fraternal rose out of the ashes like the "Phoenix" and reorganized the Klan in 1915; it happened in Georgia thanks to a former Christian preacher by the name of Colonel William Joseph Simmons. The "Invisible Empire" and "Knights of the Ku Klux Klan" was made up of native-born, white Protestant men that had somehow found a home in almost every state in America. Simmons' new organization made a target of African Americans, Jews, Catholics—all non-Protestants, liberals, union workers, teachers, and so on. They held masked marches in city streets, and many innocent American citizens were flogged, kidnapped, mutilated, and murdered—and, in some cases, the Klansmen had protection from sympathetic local officials and their cronies that would turn a "blind eye" to their sinful atrocities.

The secretive organization which first came about after the War of the Rebellion and was championed by General Robert E. Lee's foremost cavalryman, Nathan Bedford Forrest, finally met some resistance from Congress in 1871; the Force Bill was put forth to implement the Fourteenth Amendment to the Constitution of the

United States which guaranteed rights for all Americans. Once again in 1924 at the national convention of the Democratic Party a resolution was laid out which would help to stop the KKK, but it was defeated.

Hundreds of years ago on March 13, 1660, in Virginia, the first law which indicated that slavery was institutionalized was issued. Over 200 years after that on August 22, 1862, Horace Greely, editor of the New York *Tribune*, asked President Abraham Lincoln about the purpose of civil war, and he responded: "My paramount object in this struggle is to save the Union, and not either to save or destroy slavery. If I could save the Union without freeing any slaves I would do it; if I could save it by freeing all the slaves, I would do it; and if I could do it by freeing some and leaving others alone, I would do that."

Bibliography

Forrest, Nathan Bedford, *The Capture of Fort Pillow*, Battles and Leaders of the Civil War, The Century Company, 1887.

Funk & Wagnalls New Encyclopedia, Funk & Wagnalls, Inc., New York, 1979.

Grant, Ulysses S., *Preparing for the Campaigns of '64*, Battles and Leaders of the Civil War, The Century Company, 1887.

Halleck, Reuben Post, *History of Our Country*, American Book Company, 1923.

Legrand, Jacques, *Chronicle of America*, Chronicle Publications, Mount Kisco, New York, 1989.

New York *Tribune*, August 22, 1862.

Roark, James L., *The American Promise: A History of the United States*, Bedford Books, Boston, 1998.

Captain William Clarke Quantrill and the Desecration of His Bones

ONE OF THE MOST infamous raiders of the Border War and American Civil War was William Clarke Quantrill. Born in Canal Dover (Dover), Ohio, on July 31, 1837, young Quantrill was reportedly a troubled, lazy child—by some accounts. Eventually, however, he traveled to the West, experienced, and caused trouble. For many and various reasons, known only to him and maybe a few ambitious historians and history-buffs, Quantrill ended up in Missouri fighting passionately for the Southern cause—or, more likely, against his accumulated hatred of Kansas, the Jayhawkers, and the Union. Ironically, though, from such a reported slothful childhood sprang such a formidable guerrilla warrior and leader.

During his time in the Civil War, two of the most horrific attacks perpetrated by Captain Quantrill and his men occurred—first, on August 21, 1863, at Lawrence, Kansas, where over 150 unarmed men, women, and children were massacred—and, second, at Baxter Springs, Kan., on October 6, 1863, where Quantrill's band of guerrillas wiped-out about one hundred Union soldiers.

Towards the end of the war Capt. Quantrill and a number of his followers decided to travel to the East and away from Missouri and Kansas; probably fearing that their type of warfare during the Civil War would not be easily forgiven by a victorious Union. There was also some speculation that he was on his way to Washington, D.C., planning to assassinate President Abraham Lincoln. Along the way, however, they decided to spend some quality time in Kentucky pillaging the countryside. Quantrill and about 40 men, including Frank James, made their temporary headquarters at the home of James H. Wakefield who lived in Spencer County along the Salt River.

It just so happened that a young 19-year-old, Union Captain Edwin (Edward) Terrill and his band of about 50 guerrillas, had been "hot" on the trail of Quantrill and his raiders. Over a period of time, Quantrill felt comfortable enough to share with his host, Wakefield, his true identity. On May 10, 1865, as rain fell, Quantrill was napping in the haymow of Wakefield's barn, while some of his men were exercising the gift of gab and a few of the others were reportedly having a sham-battle with some corncobs. Suddenly and without warning Clark Hockensmith, one of Quantrill's men, noticed a large number of mounted, well-armed riders bearing down on them, so he let out a bloodcurdling yell to alert his comrades to the fast approaching danger. The Confederates scattered in every direction with the Federals in pursuit. In the meantime, Quantrill woke with a start and attempted to mount his horse which, unfortunately, was a borrowed one that was not used to such military-type activity and became wild and unmanageable—so he lit out on foot. Seeing their leader's predicament, Hockensmith and Dick Glasscock returned for him and while he was attempting to double-up behind one of his returning cohorts, Quantrill was shot in his spine which paralyzed him, for the most part, from the neck down; the death-dealing shot was either fired by Captain Terrill, or by John Langford who later moved to Clarinda, Iowa—the birthplace of this author.

The surprise guerrilla-style attack was over quickly. As a result of returning to help Quantrill, Hockensmith and Glasscock were also killed. Quantrill was unable to move and had also lost his right hand index finger, which had been shot off. Some of the Federals that were first to arrive snatched up Quantrill's pistols and ripped-off his boots. The old Missouri raider was still conscious and was taken back to Wakefield's house where he was laid on a couch. While this was taking place Terrill's men began to ransack the house, so Wakefield offered Captain Terrill money and whiskey to appease him and his plundering men. After Wakefield agreed to look after Quantrill and keep him there until they returned, Capt. Terrill and his men left the scene.

Since Quantrill's condition was critical, a doctor was sent for to attend to and assess his condition. After the examination Dr. McClaskey reported that his injury would be fatal. Eventually, some of Quantrill's men returned and tried to talk him in to leaving with them, but he refused. Quantrill thanked Wakefield for helping him and his men while he waited for the "grim reaper" to arrive.

The next day, Capt. Terrill and about 25 men returned to collect Quantrill. They loaded him into a horse-drawn wagon and headed for Louisville, Kentucky and the hospital located there. Quantrill clung to life for about a month but finally died. As for his band of brothers, they "holed up" in the area for a spell and eventually surrendered at Samuel's Depot (Wakefield, Kentucky) on July 26, 1865, including Frank James; the men were all paroled by General John M. Palmer—the Civil War had ended.

The body of Quantrill was buried in Louisville; however, in 1887 the Confederate Chieftain's mother and a biographer by the name of William W. Scott, reportedly, got permission to exhume the body and move it to Canal Dover (his Ohio birthplace). Things did not go as planned; because of the poor condition of the remains some of his crumbling bones had to stay in the Kentucky grave. Many of the bones, however, were taken back to Canal Dover and buried in the family plot. The only problem was that Scott, for some gruesome reason, chose to steal a few of Quantrill's bones—and skull, in a lawless desecration. In about a year, Scott donated some of his desecrated bones to the Kansas State Historical Society, and for some reason, they accepted them. When Scott passed on to the Great Unknown in 1902 he reportedly left William Elsey Connelly, a Kansas historian and author, the remaining Quantrill bones which were still in his possession; and once again, for some strange reason, Connelly also accepted the desecrated bones. In 1912, however, Connelly gave them up to the Kansas State Historical Society; apparently, some also ended up in the Kansas Museum of History in Topeka. Today, Civil War buffs, interested people, Missourians, and Kansans can visit the headstone of William Clarke Quantrill at the Confederate Memorial Cemetery in Higginsville, Missouri.

Above: Sign at the Confederate Memorial Cemetery in Higginsville, Missouri.
Below: The Confederate Memorial Cemetery in Higginsville, Mo.

Captain William Clarke Quantrill

Bibliography

Barton, O.S., *Three Years with Quantrill: A True Story by His Scout John McCorkle*, Commentary, Herman Hattaway, University of Oklahoma Press, Norman, 1992.

Connelly, William Elsey, *Quantrill and the Border Wars*, The Torch Press, Cedar Rapids, Iowa, 1910.

Enss, Chris, *Tales Behind the Tombstones: The Deaths and Burials of the Old West's Most Nefarious Outlaws, Notorious Women, and Celebrated Lawmen*, Globe Pequot Press, 2007.

Funk & Wagnalls New Encyclopedia, Funk & Wagnalls, Inc., New York, 1979.

Settle, Jr., William A., *Jesses James was his Name* or *Fact and Fiction Concerning the Careers of the Notorious James Brothers of Missouri*, Bison Books, 1966.

Colonel Samuel Colt's Multi-Shot Killing Machine

BACK IN the 19th century firearms were an important part of survival and everyday life, for hunting and protection—there was less law enforcement at the time. Another place that weaponry was in constant use was by soldiers, irregulars, and outlaws who left a trail of death and destruction behind them wherever they went. The tread only increased with time and mechanical breakthroughs. There was one man who did more than just about anyone else to change a less deadly single-shot weapon into a multi-shot killing machine.

Samuel Colt was born in Hartford, Connecticut on July 19, 1814. His father was in the textile industry spinning and weaving silk. Samuel worked in the mill and went to school but also liked to tinker around with mechanical things, such as his father's gun collection. When Samuel was about 16-years-old he ran away from home and joined the crew as a seaman on the merchant ship, *Corlo*, which was bound for Calcutta, India.

While on this voyage, however, Colt was inspired to carve-out of wood his first model of a multi-shot handgun. According to one report: "...it was while handling the old bell-mouthed, brass-barreled, clubby firearms carried by the ship as a defense against pirates, that he conceived the idea of the revolving cylinder containing chambers to be discharged through a single barrel."

Other stories contend that Colt envisioned his bloody invention while watching the ship's revolving waterwheel; still others maintain that the revelation may have dawned on Colt while seeing

CURIOUS AND UNUSUAL CIVIL WAR STORIES

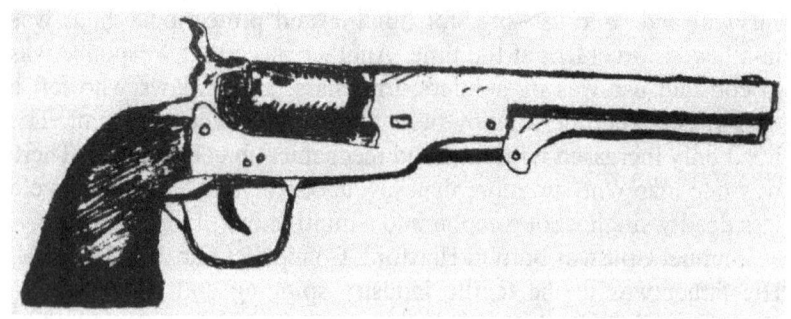

Author's sketch of the Colt revolver.

the ship's capstan being used. Regardless of how he received his mechanical wizardry and inspiration for the revolver, he began to make plans for its construction—first, by showing his father the crude wooden model that he had whittled-out with the hopes of gaining his help to put it into production; however, the idea was met with disappointment. Some people even went so far as to ridicule the notion thinking that it was "wholly impracticable and useless."

Colt did not give up on the idea. In 1836, at the age of 22, after obtaining a patent, he established the "Patent Arms Company" in Paterson, New Jersey. He believed that his revolver could greatly enhance the firepower of military infantrymen—who, at the time, had to reload after each discharge. Colt's first financial success was a result of arms sales during the Florida Seminole Removals—a 7-year conflict that harvested the lives of about half the Seminole tribe. Colt's cargo of revolvers to Florida were well-received by the soldiers and their use brought slaughter and death to the Indians who "could not understand how so many shots were fired without reloading, and they pronounced the revolver 'big medicine.' "

As time went on and he was unable to gain ground in the industry, the company went out of business a few years later in 1842. Not to be detoured from his ambitions altogether, Colt began to invent submarine mines (torpedoes) and also ran a telegraph business where he developed the first underwater cable.

In 1843 after a Texas naval victory over the Mexican fleet, Colt designed his most famous revolver, the "navy" pistol which sported an engraving of the historic battle on the weapon's cylinder. The Colt navy revolver went on to become one of the most popular firearms ever created, and was used by both the North and the South during the American Civil War as well as by bushwhackers, guerrillas, and others.

It was said that William Clarke Quantrill and his men never carried anything else but the Colt navy revolver. The Kansas author, William Elsey Connelly, wrote about its use by guerrillas and said: "Every guerrilla carried two revolvers, most of them carried four, and many carried six, some even eight. They could fire from a revolver in each hand at the same time. The aim was never by sighting along the pistol-barrel, but by intuition, judgment. The pistol was brought to the mark and fired instantly, apparently without a care, at random. But the ball rarely missed the mark—the center. Many a guerrilla could hit a mark to both the right and the left with shots fired at the same instant from each hand.

"No more terrifying object ever came down a street than a mounted guerrilla wild for blood...When a town was filled with such men bent on death, terror ensued, reason and judgment fled, and hell yawned."

At the outbreak of the Mexican War, which was waged from 1846-1848, Colt received a government contract to mass-produce his weapons; the U.S. government purchased 1,000 Colt revolvers. By 1852, Colt had built an armory in Hartford, Conn. where he produced weapons and other machinery used in the manufacturing of firearms. Colt is known for being one of the first to introduce parts that are interchangeable in manufacturing.

Colonel Samuel Colt is also remembered for saying that "his pistol would make all men equal." Colt died in 1862—three years before the end of the Civil War. The inventor would not live to see the total death-toll of that bloody, homegrown war—about 700,000 souls and many more maimed and crippled. Colt's multi-shot firearms would, not only contribute to that deadly number, but would go on to become one of the 19th century's most sought after products—a curious and unusual fact of American Civil War history and culture.

Bibliography

Connelly, William Elsey, *Quantrill and the Border Wars*, Cedar Rapids, Iowa, The Torch Press, 1910.

Jackson, Rex T., *Colonel Samuel Colt, A 19th Century American Inventor*, Vol. 8, No. 3, The Ozarks Reader Magazine, Neosho, Missouri, 2011.

Legrand, Jacques, *Chronicle of America*, Chronicle Publications, Mount Kisco, New York, 1989.

Additional Illustrations

Additional illustrations

Above: Officer and troop quarters at Fort Scott, Kansas.
Below: Barracks storehouse at Fort Scott, Kansas.

Blockhouse at Fort Scott, Kansas.

Mural of guerrilla warfare in Nevada, Missouri.

Mural of soldiers burning buildings at Nevada, Missouri.

Above: Kendrick House in Carthage, Missouri.
Below: Stone marker at the Kendrick House in Carthage, Missouri.

Historic plaque on the Kendrick House in Carthage, Missouri.

Civil War reenactment in Carthage, Missouri.

Fort Gibson, Oklahoma

Reproduction grand staircase of the *Titanic* at the "Titanic Museum" in Branson, Missouri.

Index

Index

A

Abel, Annie Heloise, 13, 25
Abilene *Chronicle*, 33
Abilene, Kansas, 33
Albany, Missouri, 78
Americus, Georgia, 67
Anderson, Janie, 75
Anderson, Josephine, 75
Anderson, Molly, 75
Anderson, Oliver, 15, 17, 18, 19, 20
Andersonville Prison, 65, 66, 67, 68, 69, 70
Anderson, William T. ("Bloody Bill") 73, 74, 75, 76, 77, 78, 79
Annonay, France, 119
Atlantic, 119, 120
Auschwitz Prison, 65
Austin, Texas, 33

B

Baker, Frances, 77
Baker, Sylvester Marion, 77
Barnes, James, 9
Barton, Clarissa ("Clara") Harlowe, 101, 103
Barton, O.S., 26
Beauregard, Pierre Gustave T., 81, 101
Becknell, William, 39
Belfast, Ireland, 111
Benjamin, Charles F., 7
Benton, Thomas Hart, 40, 129
Bentonville, Arkansas, 98, 108
Bent, William, 38
Berlin, Germany, 67
Bledsoe, Hiram Miller, 82, 85, 86
Boggsville, Colorado, 41
Bonham, Texas, 33
Boone, Daniel, 37
Boonville, Missouri, 82
Bosque Redondo, New Mexico, 41
Branson, Missouri, 112
Britton, Wiley, 25, 77
Buchenwald Prison, 65
Buffalo, New York, 119
Bulger, M.J., 9
Burnside, Ambrose E., 7

Index

C

Cairo, 91
Calcutta, India, 155
Campbell, J.M., 9
Camp Morton, 109
Canal Dover, Ohio, 149, 151
Carondelet, 91, 93
Carson, Christopher Houston ("Kit"), 39, 40, 41, 131
Carson City, Nevada, 41
Carthage, Missouri, 3, 81, 82, 83, 84, 85, 86, 87, 88
Centralia, Missouri, 76, 77
Chamberlain, Joshua L., 9, 10
Chandler, D.T., 69
Charleston, South Carolina, 81, 129
Cherbourg, France, 55, 58, 59, 60
Chiluahua, Mexico, 40, 85
Cincinnati, 91, 93, 94
Cincinnati, Ohio, 34, 120
Clara Bell, 20
Clarinda, Iowa, 150
Cody, William F. ("Buffalo Bill"), 34
Coe, Phil, 33
Coffin, William G., 141
Colt, Samuel, 155, 156, 157, 158
Connelley, William Elsey, 76, 151, 157
Cooper, Douglas H., 139

Corlo, 155
Cox, Samuel P., 78
Cracow, Poland, 65
CSS Alabama, 55, 56, 57, 59, 60, 61

D

Danville, Missouri, 77
Deadwood, South Dakota, 34, 35
Davis, Jefferson C., 20, 45
Davis, Jefferson (President), 49, 79, 81, 101, 132
Dawson, Thomas, 87
Deerfield, Missouri, 14, 45
Doniphan, Alexander W., 85
Dover, Ohio, 149

E

Eads, James Buchanan, 91, 92
Edmonds, Sarah Emma, 102, 103, 104
Ewing, Thomas, 75

F

Flagstaff, Arizona, 40

Index

Flushing, Holland, 56
Foote, Andrew H., 91, 94
Forrest, Nathan Bedford, 143, 145, 146
Fort Bent, 38
Fort Donelson, 94
Fort Gibson, Oklahoma, 23, 137
Fort Heiman, 93
Fort Henry, 91, 93, 94
Fort Leavenworth, Kansas, 140
Fort Pillow, Tennessee, 143, 145
Fort Riley, Kansas, 32
Fort Scott, Kansas, 14, 23, 24, 25, 26, 27, 45, 46
Fort Smith *New Era*, 78
Fort Wayne (Oklahoma), 23
Foster, Joe, 33
Franklin, Missouri, 39
Fremont, John C., 20, 45, 129, 131, 132, 133, 134
Fyan, Robert W., 32

G

Gager, O.A., 119
Gallatin, Missouri, 78
Gettysburg, Pennsylvania, 7, 8, 10, 52
Glasscock, Dick, 150
Goss, Warren Lee, 2, 68
Gracie, Archibald, 111, 113
Granby, Missouri, 34
Grant, Ulysses S., 132, 145
Greely, Horace, 147

Greenhow, Rose O'Neal, 100, 101

H

Halleck, Reuben Post, 51, 52, 57, 62
Hancock, Winfield Scott, 132
Hunt, Henry J., 8
Hardie, James A., 8
Hardin, John Wesley, 33
Harris, Gen., 18
Hartford, Connecticut, 155, 158
Hatteras, 58
Hawks, Francis Lister, 37
Hays City, Kansas, 32, 33
Hays City *Sentinel*, 33
Henderson, New York, 120
Hickok, James Butler ("Wild Bill"), 30, 31, 32, 33, 34, 35
Higginsville, Missouri, 151, 152
Hitler, Adolf, 65
Hockensmith, Clark, 150
Hooker, Joseph, 7, 8
Hopoeithleyohola, 139, 140, 141
Huntsville, Missouri, 73
Hyde, William, 119

I

Independence, Missouri, 38

Index

Indianola, 125, 126
Intrepid, 121

J

Jackson, Claiborne Fox, 82, 86
Jackson, James W., 9
James, Frank, 77, 79
James, Jesse, 77, 79
Jane, Calamity, 35
Jaramillo, Maria Josefa, 40, 41
Jefferson City, Missouri, 45
Jennison, Charles R., 26
Johnson, A. Sidney, 94
Johnston, A.V.E., 77
Jone, Joseph, 69

K

Kansas City, Missouri, 75
Kearney, Stephen W., 26, 132
Kell, John McIntosh, 58, 59
Kendrick, William B., 87
Kerr, Charity, 75
Kerr, Josephine, 75
Killian, Joe, 34
Knottingley Yorkshire, England, 33

L

Lake, Agnes Thatcher, 34
LaMountain, John, 119, 120
Lancashire, England, 60
Lancaster, John, 60
Lane, James H., 14, 25, 26, 45, 46, 140
Lawrence, Kansas, 75, 149
Leavenworth, Kansas, 23, 26, 140
Lee, Robert E., 7, 8, 11, 132, 146
Lexington, Missouri, 13, 14, 16, 20, 21, 45, 116
Libby Prison, 70
Lincoln, Abraham (President), 7, 11, 26, 27, 49, 50, 71, 133, 140, 147, 149
London, England, 50
Longstreet, James, 120, 122, 132
Louisville, 91
Louisville, Kentucky, 151
Lowe, Thaddeus S.C., 120, 121
Lyon, Nathaniel, 81, 82, 86, 87, 99, 100

M

Making-Our-Road, 40
Manet, Edouard, 58

Index

Marshall, James, 131
Maysville, Arkansas, 23
McBride, Gen., 18
McCall, Jack, 34
McClellan, George, 7, 141
McCulloch, Ben, 87
McDowell, Irvin, 101
McKenna, Charles F., 4
McKnight, Robert, 39
Meade, George G., 7, 8, 11
Melcher, H.S., 10
Merrimac, 59, 60, 62, 91, 94
Messman, Agnes Louise, 34
Monitor, 59, 60, 62, 91, 94
Montgolfer, Jacques Etienne, 119
Montgolfer, Joseph Michel, 119
Montgomery, James, 25, 45
Mound City, 91
Mound City, Illinois, 91
Mountain Home, Arkansas, 31
Mulligan, James A., 14, 16, 17, 18, 19, 20

N

Nevada, Missouri 14, 43, 44
Newtonia, Missouri, 31
New York *Tribune*, 52, 147
New York *Times*, 81
Niagara Falls, 119
Nicholasville, Kentucky, 15
Nicollet, Joseph Nicolas, 129
Nova Scotia, Canada, 103

O

Oakley, Annie, 34
Oakley, Daniel, 1
Opothleyoholo, 139
Orrick, Missouri, 78
Oxford, Massachusetts, 101

P

Parkman, Francis, 38
Parsons, Monroe M., 45, 100
Paterson, New Jersey, 157
Peabody, Col., 16
Pea Ridge, Arkansas, 31, 100
Phelps, Mary Whitney, 99
Philadelphia, Pennsylvania, 119
Phoenix, Arizona, 40
Picket, George, 11
Pittsburgh, 91
Polk, James Knox (President), 133
Pope, John, 7
Porter, David D., 126
Porter, W.D., 93
Portland, Maine, 99
Price, Sterling, 14, 16, 17, 18, 19, 20, 43, 45, 46, 79, 82, 87, 100
Pulaski, Tennessee, 145
Putman, George H., 69

Index

Q

Quantrill, William Clarke, 26, 75, 76, 149, 150, 151, 157
Queen of the West, 125, 126
Queenstown (Cobh), Ireland, 111

R

Rains, James S., 14, 45, 78, 82, 84
Rankin, Sinnet, 87
Richmond, Missouri, 78, 79
Richmond, Virginia, 61, 67, 70, 81, 101 115, 122
Roberson, John, 34

S

Sacramento, California, 131
San Antonio, Texas, 33
San Diego, California, 40
Santa Fe, New Mexico, 38, 39
Savannah, Georgia, 129
Scott, William W., 151
Scott, Winfield, 23
Seelye, Linus, 104
Semmes, Raphael, 55, 56, 58, 59, 60, 61

Shelby, Joseph Orville, 82, 84
Sherman, William Tecumseh, 13, 132
Sigel, Franz, 82, 84, 85, 86, 87
Simmons, William Joseph, 146
Sitting Bull, 34
Snead, Thomas L., 84
Snow, George C., 141
Springfield, Missouri, 14, 30, 32, 43, 85, 99, 100
Springfield *Republican*, 31
Stemble, R.N., 91
St. Louis, 91, 93
St. Louis, Missouri, 20, 37, 40, 45, 81, 91, 119, 131, 132, 133
Straus, Isidor, 111
Sturgis, Thomas, 70
Sutter's Fort, 131
Sykes, George, 9

T

Taos, New Mexico, 39, 41
Taylor, Zachary, 132
Terrill, Edwin, 150, 151
Thatcher, William Lake, 34
Thompson, Ben, 33
Tilghman, Lloyd, 93, 94
Titanic (RMS), 55, 111, 112
Todd, George, 77, 79
Troy Grove, Illinois, 31
Tutt, Davis K., 31, 32

Index

U

USS Kearsarge, 55, 56, 58, 59, 60, 61

V

Vicksburg, Mississippi, 126
Virginia, 59, 91

W

Waa-Nibe, 40
Wakefield, James H., 150
Wakefield, Kentucky, 151
Walke, Henry, 93
Washington, George, 10
Watts, Hamp B., 73
Webb, 125, 126
Webb, W.L., 16, 19, 86
Weed, Gen., 26
Winder, John H., 67
Winslow, John Aucrum, 56, 58
Wirz, Heinrich Hartmann, 67, 69, 70, 71
Wise, John, 119, 120
Wood, John, 60
Wyeth, John Allan, 99

X

Y

Yellville, Arkansas, 31

Z

Zurich, Switzerland, 67

181

Other Books by Rex T. Jackson

The Sultana Saga: The Titanic of the Mississippi
James B. Eads: The Civil War Ironclads and His Mississippi
A Trail of Tears: The American Indian in the Civil War
Traces of Ozarks Past: Outlaws, Icons, and Memorable Events
Monumental Tales from the Ozarks
Notable Persons and Places in Missouri's History
Timeless Stories of the West: Mountaineers, Miners, and Indians

About the Author

Rex T. Jackson's work has appeared in a number of publications, like *The Ozarks Mountaineer, Blue and Gray, Good Old Days, Ancient American, Capper's Weekly, Back Home, The Ozarks Reader, Route 66 Magazine* and others. He became a staff member of *The Ozarks Mountaineer* (based in the Branson, Missouri area) which began in 1952 and ended in 2012, and eventually founded *The Ozarks Reader Regional Magazine* and served as publisher and editor from 2004 through 2012.

www.ingramcontent.com/pod-product-compliance
Lightning Source LLC
Chambersburg PA
CBHW071419160426
43195CB00013B/1748